T0296320

SOCIETY FOR EXPERIMENTAL BIOLOGY
SEMINAR SERIES: 31

PLANT CANOPIES:
THEIR GROWTH, FORM AND FUNCTION

SOCIETY FOR EXPERIMENTAL BIOLOGY SEMINAR SERIES

A series of multi-author volumes developed from seminars held by the Society for Experimental Biology. Each volume serves not only as an introductory review of a specific topic, but also introduces the reader to experimental evidence to support the theories and principles discussed, and points the way to new research.

PLANT CANOPIES: THEIR GROWTH, FORM AND FUNCTION

Edited by

G. Russell,

University of Edinburgh

B. Marshall,

Scottish Crop Research Institute
and

P.G. Jarvis,

Professor of Forestry and Natural Resources, University of Edinburgh

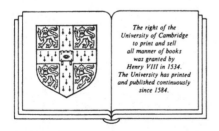

The right of the
University of Cambridge
to print and sell
all manner of books
was granted by
Henry VIII in 1534.
The University has printed
and published continuously
since 1584.

CAMBRIDGE UNIVERSITY PRESS

Cambridge

New York   Port Chester   Melbourne   Sydney

Published by the Press Syndicate of the University of Cambridge
The Pitt Building, Trumpington Street, Cambridge CB2 1RP
40 West 20th Street, New York, NY 10011, USA
10 Stamford Road, Oakleigh, Melbourne 3166, Australia

First published 1989
First paperback edition 1990

British Library cataloguing in publication data

Plant canopies: their growth, form and
function – (Society for Experimental
Biology seminar series; 31).
1. Leaves
I. Russell, G.   II. Marshall, B.
III. Jarvis, P.G.   IV. Series
581.4'97   QH649

Library of Congress cataloguing in publication data

Plant canopies: their growth, form, and function/edited by G.
Russell, B. Marshall, and P.G. Jarvis
p.   cm. – (Society for Experimental Biology seminar series: 31)
Based on seminars held during the March 1986 meeting of the
Society for Experimental Biology at Nottingham: organized by the
Environmental Physiology Group
Includes index.
ISBN 0 521 32838 1
1. Plant canopies – Congresses.   I. Russell, G. (Graham)
II. Marshall, B. (Bruce)   III. Jarvis, P.G. IV. Society for
Experimental Biology (Great Britain). Meeting (1986: Nottingham.
Nottinghamshire) V. Society for Experimental Biology (Great
Britain). Environmental Physiology Group. VI. Series: Seminar
series (Society for Experimental Biology (Great Britain)): 31.
QK924.5.P53 1989
581.5–dc19   87–32902

ISBN 0 521 32838 1 hardback
ISBN 0 521 39563 1 paperback

Transferred to digital printing 2003

DQ

CONTENTS

CONTRIBUTORS

G.S. Campbell
Dept. of Agronomy and Soils, Washington State University, Pullman, WA 99164, USA.

J.R. Ehleringer
Dept. of Biology, University of Utah, Salt Lake City, Utah 84112, USA.

I.N. Forseth
Dept. of Botany, University of Maryland, College Park, MA 20742, USA.

J. Goudriaan
Dept. of Theoretical Production Ecology, Agricultural University, Wageningen, The Netherlands.

J.L. Harper
Unit of Plant Population Biology, University College of North Wales, Bangor, Gwynedd LL57 2UW, UK.

P.G. Jarvis
Institute of Ecology and Resource Management, University of Edinburgh, Edinburgh EH9 3JU, UK.

H. van Keulen
Centre for Agrobiological Research, Wageningen, The Netherlands.

K.G. McNaughton
Plant Physiology Division, DSIR, Palmerston North, New Zealand.

J.L. Monteith
Dept. of Physiology and Environmental Studies, University of Nottingham, Sutton Bonington, Loughborough LE12 5RD, UK.
(Present address: ICRISAT, Patancheru P.O., Andhra Pradesh 502 324, India.)

J.M. Norman
Dept. of Soil Science, University of Wisconsin, Madison, WI 53706, USA.

J.R. Porter
Dept. of Agricultural Sciences, University of Bristol, Long Ashton Research Station, Bristol BS18 9AF, UK.

M.R. Raupach
CSIRO Division of Environmental Mechanics, GPO Box 821, Canberra, ACT 2601, Australia.

G. Russell
Institute of Ecology and Resource Management, University of Edinburgh, Edinburgh EH9 3JG, UK.

N.G. Seligman
Agricultural Research Organisation, Bet Dagan, Israel.

PREFACE

Recent advances in modelling plant stands have emphasised the importance of the structural and functional properties of plant canopies, as distinct from those of the constituent parts. In response to proposals made following the 1984 meeting on the 'control of leaf growth', which resulted in Seminar Series Publication 27, the Environmental Physiology Group held a series of sessions on plant canopies during the March 1986 meeting of the Society for Experimental Biology at Nottingham. All the invited speakers at these sessions have contributed chapters to this volume either individually or with collaborators.

Chapters have been included on all the major processes occurring in canopies, although there has been space neither for consideration of the manipulation of canopies by chemical or genetical means, nor for discussion of the canopy as habitat for micro-organisms, insects or vertebrates. A policy decision was made at an early stage of planning to encourage authors to look at a diverse range of canopy types and geographical distribution in order to avoid any bias introduced by, for example, considering only temperate zone cereal crops. The reader can decide how successful this policy has been. Some omissions represent genuine areas of ignorance, but it is a matter of regret that space was not available to allow consideration of stands of mixed species either in agricultural intercropping systems or in natural communities.

It is a pleasure to acknowledge the financial and other support of the Environmental Physiology Group, the Association of Applied Biologists and the British Ecological Society.

I would like to record the contributors' co-operation during the meeting and to thank them for all the time they and their collaborators devoted to preparing and revising their manuscripts. The editors would also like to acknowledge the assistance of colleagues who gave advice and refereed the chapters.

Graham Russell
Edinburgh

G.S. CAMPBELL AND J.M. NORMAN

1. The description and measurement of plant canopy structure

Introduction

Plant canopy structure is the spatial arrangement of the above-ground organs of plants in a plant community. Leaves and other photosynthetic organs on a plant serve both as solar energy collectors and as exchangers for gases. Stems and branches support these exchange surfaces in such a way that radiative and convective exchange can occur in an efficient manner. Canopy structure affects radiative and convective exchange of the plant community, so information about canopy structure is necessary for modelling these processes.

In addition to considering how canopy structure and environment interact to affect the processes occurring in the plant community, the influence of the canopy on the environment should also be considered. The presence and structure of a canopy exert a major influence on the temperature, vapour concentration, and radiation regime in the plant environment. Interception and transmission of precipitation are also affected, as are soil temperature and soil heat flow. Canopy structure can therefore be important in determining the physical environment of other organisms within the plant community, and can strongly influence their success or failure. Plant canopy structure can indirectly affect such processes as photosynthesis, transpiration, cell enlargement, infection by pathogens, growth and multiplication of insects, photomorphogenesis, and competition between species in a plant community. The indirect influence on soil moisture and temperature can also affect root growth, evaporative water losses from the soil, residue decomposition and other soil microbial processes.

A complete and accurate description of a canopy would require the specification of the position, size and orientation of each element of surface in the canopy. Such a description is clearly impossible to obtain, except for very simple canopies, so that data needs in terms of specific applications must be carefully considered. Canopy properties are generally described statistically as appropriate space or time averages. In some cases additional statistical parameters are needed for an adequate description of the canopy.

Canopies vary on spatial scales ranging from millimetres to kilometres, and on time scales ranging from milliseconds to decades. The description of this variation is an important part of understanding and using canopy structure information. Consideration of variation in structure can be useful in recognizing patterns which may exist, and using these patterns to maximize sampling efficiency or minimize sampling errors.

Application of the principle of least work (Monteith, 1985) is particularly appropriate to measurements of canopy structure. Because it is possible to invest a lot of time and effort in measurements of canopy structure, it is particularly desirable to determine data requirements before an extensive measurement programme is undertaken.

Phytometric characteristics of plant canopies

The characterisation of plant canopies using various statistical parameters has been presented in considerable detail by Ross (1981). The material presented here is intended to be a brief summary. The reader is referred to the original work for additional detail.

Ross (1981) recommends that descriptions of plant canopies should include measurements at four levels of organisation: individual organs, the whole plant, the pure stand, and the plant community. Each higher level of organisation is intended to include elements from the next lower level, and to add parameters of its own.

At the individual organ level, parameters such as typical length, width, area, dry mass, specific water content, and radiative properties of phytoelements are measured. Whole plants are often symmetric, and have outlines which can be represented by some geometric shape. The parameters which describe the geometric shape are therefore useful as parameters for describing average characteristics of individual plants. Ellipsoidal shapes have been suggested as good approximations to plant outlines (Charles–Edwards & Thornley, 1973; Mann, Curry & Sharpe, 1979; Norman & Welles, 1983). Plants which cannot be represented by a complete ellipsoid can often be represented by a truncated ellipsoid. In addition to plant height and other parameters which relate to the overall geometric shape of the plant, it may be useful to record stem diameter at one or more locations, total number of leaves per plant, number of nodes per plant, number of living leaves per plant, numbers of stems and reproductive organs per plant, and spatial distribution of organs within the plant outline.

In an attempt to maximise return for a given sampling effort, Ross (1981) suggested a two-stage sampling process in which primary statistical characteristics such as plant height, height of the top and bottom of the foliage canopy, stem height and diameter, number of leaves (where possible) and number of living leaves (where possible) are determined on an initial sample of 150–300 plants. These primary characteristics are then examined to select 15–30 plants to be analysed in greater detail to determine average characteristics of individual organs, spatial locations of organs, and orientation of surfaces. If it is not possible to determine spatial locations of organs within the plant envelope, parameters for simple models of foliage distribution should be obtained. Mann *et al.* (1979) suggested three possible idealised distribution functions. The uniform, the quadratic, and the truncated normal. The uniform distribution is based on the assumption that the probability of finding an element at any location within the plant envelope is independent of position. The other two distribution functions assume a higher density of foliage near the centre of the envelope. Norman & Welles (1983) assumed a uniform density of foliage within the ellipsoidal envelopes within which individual plants are contained, but allowed for the possibility that ellipsoids with different densities could be placed concentrically. Variations of area density within a given plant envelope were described by specifying the dimensions of the various ellipsoidal shells and the average foliage density within each shell.

At the pure stand and plant community levels of organisation, Ross (1981) suggested four types of plant dispersion: regular, semi-regular, random, and clumped. Regular dispersion results when plants are located at the vertices of a regular parallelogram. An example of this would be an orchard planting, or a square or hexagonally sown crop. Semi-regular dispersion results when plants are in rows, but spacing within the row is random, as in many agricultural crops. In random dispersion, there is an equal probability of finding a plant at any location, and with clumped dispersion the probability of finding a plant in a given location is related to the presence or absence of plants in the surrounding area.

A description of canopy organisation at the pure stand or plant community level requires, at least, a measurement of the plant population density, i.e. the number of each species of plant per unit area. For regular or semi-regular dispersion, plant or row spacings are needed and for regular dispersion, angles of the vertices of the parallelogram should also be determined. For random dispersion, only the plant population density is relevant, while for clumped dispersion, it may be possible to assume random dispersion within clumps, and define the size and distribution of the clumps.

As plants grow, they may begin to overlap so that it is difficult to discern the outline of a particular plant or row of plants. The time at which this occurs is termed 'canopy closure'. After canopy closure, radiative exchange and heat and mass transfer processes can be treated using one-dimensional theory. Large plant communities, where the horizontal dimensions are much larger than the vertical dimension, can also be treated as one-dimensional. A one-dimensional model allows dramatic simplification of the convective and radiative exchange processes. It may then be assumed that the phytoelements are randomly distributed in space (rather than within the plant envelope) or grouped around shoots which may themselves be randomly distributed in space.

Canopies which can be modelled as a series of horizontal layers, using one-dimensional models, are important in many agricultural and forest applications, and much of the following analysis will deal with this simplified canopy type. Such canopies are often described in terms of two parameters, the average area density of component j, $\mu_j(z)$ (m^2m^{-3}), and the angle distribution function of component j, $g(z,r_j)$. The index j is intended to apply to leaves (l), stems (s), and reproductive parts (f) of the plant. The variable z, represents height in the canopy, and r_j represents the direction of a normal to the canopy element (i.e. azimuth, ϕ_j, and inclination θ_j, angles). The function, $g(z,r_j)$ represents the probability of a normal to a canopy element falling within an angle increment, $d\theta,d\phi$. It is normalised so that the integral of $g(z,r_j)$ over all angles in a hemisphere is unity.

Integrals of these parameters are often used. The downward cumulative area index of component j in a canopy is

$$L_j(z) = \int_z^h \mu_j(z)\, dz \qquad (1)$$

where h is the height of the top of the canopy, and z the height from the ground. The leaf area index of a canopy (L_o) is the total area of leaves above unit area of soil, and is given by eqn (1) when $j = l$ and the lower limit of integration is equal to zero.

The integral of the angle distribution function $g(z,r_j)$, is the canopy extinction coefficient for a beam of radiation. This integral can be thought of as the average projected area of canopy elements or the ratio of projected to actual element area. Ross & Nilson (1965) define a G-function, which is the average projection of canopy elements onto a surface normal to the direction of the projection. If the projection zenith angle is θ and azimuth angle ϕ, then the G-function is calculated from the weighted integral of $g(z,r_j)$ over the hemisphere:

$$G(z,r) = \frac{1}{2\pi} \iint g_j(z,r_j)\, |\cos(r_j,r)|\, d\theta_j d\phi_j , \qquad (2)$$

where

$$\cos (r_j,r) = \cos \theta_j \cos \theta + \sin \theta_j \sin \theta \cos(\theta - \theta_j) , \qquad (3)$$

θ_j is the inclination angle of the canopy elements (angle between the vertical and a normal to the element) and θ_j is the azimuth angle of the normal to the foliage element. The integral is taken over azimuth angles from 0 to 2π and inclination angles from 0 to $\pi/2$.

A different extinction coefficient, the K-function, has been used by a number of authors (Warren–Wilson, 1965, 1967; Anderson, 1966, 1970). It is the average projected area of canopy elements when they are projected onto a horizontal plane. It is related to the G-function by

$$K(z,r) = G(z,r)/\cos \theta . \qquad (4)$$

Simplifications and idealizations

Having established some of the fundamental parameters that can be used to characterise canopy structure, attention is now given to simplifications and assumptions that reduce the number of measurements needed to describe the canopy.

The equations presented in this section will be useful in the analysis of the methods of measurement which will be covered later. We will begin by assuming a closed, horizontally homogeneous canopy with canopy elements randomly spaced in the horizontal. We will therefore be concerned only with the total area and vertical distribution of canopy elements, and the angular distribution of canopy elements. We will assume azimuthal symmetry of the plants, since measurements (Ross, 1981; Lemeur, 1973) indicate that canopies often approximate to this.

Many of the models that are used for calculating radiant energy interception by canopies require information only on area index and angle distribution, but models for the turbulent exchanges of heat and mass, and calculations of the size of penumbra also require a knowledge of the vertical distribution of area within the canopy. Ross (1981) presented a number of examples of area density functions, $u(z)$, for various canopies. Norman (1979) and Pereira & Shaw (1980) modelled $u(z)$ as a simple triangle (Figure 1.1). In such a case, the area density is given by

$$u(z) = u_m (z - z_l)/(z_m - z_l), \quad z_l \leqq z \leqq z_m,$$

$$u(z) = u_m (h - z)/(h - z_m), \quad z_m \leqq z \leqq h, \tag{5}$$

with $u(z)$ assumed zero outside this range. The maximum leaf area density, μ_m is calculated from

$$\mu_m = 2L_0/(h - z_l). \tag{6}$$

In eqns. (5) and (6), h is the canopy height, z_l is the lower boundary of the canopy, z_m is the height of maximum leaf area density, and L_0 is the leaf area index of the canopy.

The relationship between downward cumulative area index and height for such a triangular area density distribution is found by integration of eqn (1). For the triangular distribution assumed in eqn (5) the solution is

$$L/L_0 = (1 - z/h)^2/[(1 - z_m/h)(1 - z_l/h)], \quad z_m \leqq z \leqq h,$$

$$L/L_0 = 1 - (z - z_l)^2/[(h - z_l)(z_m - z_l)], \quad z_l \leqq z \leqq z_m. \tag{7}$$

These equations allow one to describe the spatial distribution of canopy elements using four easily measured parameters: h, z_m, z_l, and L_0. Fig. 1.1 compares measured and predicted $u(z)$ and $L(z)$ for a maize canopy and indicates that these simple descriptions are adequate for many of the purposes for which spatial distribution information is needed.

Idealized leaf angle distribution functions have been widely used to approximate actual leaf angle distributions. Several formulae have been given for constant leaf inclination angles (but randomly distributed azimuthal angles). If all the leaves are

6   G.S. CAMPBELL AND J.M. NORMAN

inclined at a constant angle, θ_0, then the angle distribution function is given by (Ross, 1981):

$$g(\theta_j) = \delta(\theta_j - \theta_0) \sin \theta_j ,\tag{8}$$

where $\delta(\theta_j - \theta_0)$ is the Dirac delta function, the value of which is unity when $\theta_j = \theta_0$ and zero otherwise. A horizontal distribution results when $\theta_0 = 0$, a vertical or cylindrical distribution when $\theta_0 = \pi/2$, and a conical distribution when θ_0 is between these values. It is useful to think of the distribution of leaf angles in a canopy as being similar to the distributions of areas on various geometric objects. For example, if the surface area of a cone, cylinder or horizontal plane were divided into small elements, and the angle distribution of normals to the elements were determined, the angle distribution of these normals would form a conical, cylindrical or horizontal distribution function.

Another useful distribution function is the spherical, or uniform distribution. The distribution of angles in a canopy with a spherical leaf angle distribution is similar to the distribution of angles for small surface elements of a sphere. The angle distribution function is

$$g(\theta_j) = \sin \theta_j .\tag{9}$$

With the exception of the spherical distribution, the distribution functions described so far are discontinuous, and not at all representative of real canopies. Lemeur (1973) suggested simulating real canopies as weighted sums of conical canopies having a range of inclination angles. This has been useful in providing approximations to canopy angle distributions, but requires many parameters to quantify the inclination angle distribution. A more general form of the spherical distribution function, which is continuous over the entire range of leaf angles, but which has horizontal or vertical

Fig. 1.1. Triangular distribution of canopy area density and the resulting leaf area index distribution. Data points are for a maize canopy, and are taken from Pereira & Shaw (1980).

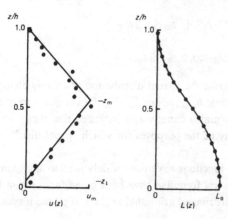

tendencies would be useful. The ellipsoidal distribution (Campbell, 1986) provides such a function. It is based on the assumption that the leaf angles in a canopy are distributed like the angles of normals to small area elements on the surface of an ellipsoid. A single parameter, $x = b/a$, is required to describe the shape of the distribution; b is the horizontal semi-axis of the ellipsoid, and a is the vertical semi-axis. When $x = 1$, the ellipsoidal distribution becomes the spherical distribution given by eqn (9). When $x > 1$ (oblate spheroid),

$$g(\theta_j) = \frac{2\,x^2\,\sin\,\theta_j}{A_1(\cos^2\,\theta_j + x^2\,\sin^2\,\theta_j)^2} \tag{10}$$

and when $x < 1$ (prolate spheroid),

$$g(\theta_j) = \frac{2\,x^2\,\sin\,\theta_j}{A_2(\cos^2\,\theta_j + x^2\,\sin^2\,\theta_j)^2}. \tag{11}$$

Here,

$$A_1 = 1 + \frac{\ln[(1+\varepsilon_1)/(1-\varepsilon_1)]}{2\varepsilon_1 x^2} \quad,\quad \varepsilon_1 = (1-x^{-2})^{1/2} \tag{12}$$

and

$$A_2 = 1 + (\sin^{-1}\varepsilon_2)/(x\varepsilon_2) \quad,\quad \varepsilon_2 = (1-x^2)^{1/2}. \tag{13}$$

Figure 1.2 shows examples of the ellipsoidal angle distribution function for several values of x.

For most purposes the extinction coefficients G, or K (eqns (2) or (4)) are more useful than the leaf angle distribution functions. These may be obtained by integrating the distribution functions using eqn (2), but are often easier to derive by considering the projected areas of solids having the angle distributions for the given distribution function (Monteith & Unsworth, 1990). Thus, for a horizontal distribution, G is the

Fig. 1.2. Ellipsoidal inclination angle distributions for several values of x which are typical of plant canopies.

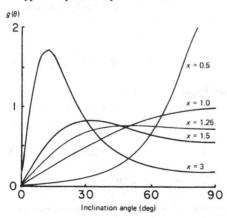

ratio of projected area of a horizontal plane to surface area of the plane, so $G = \cos\theta$. From eqn (4), $K = 1$ for the horizontal distribution. The G and K functions for other angle distributions are given in Table 1.1. Extinction coefficients are plotted as functions of zenith angle for several angle distributions in Figs. 1.3 and 1.4.

The assumption that element normals have random azimuthal distribution is in error for species with heliotropic leaves. Shell & Lang (1975) suggest the use of the von Mises probability density function to model leaf angle distributions for such canopies. Mann et al. (1979) propose a much simpler, but less realistic formula based on the assumption that all heliotropic leaves maintain a constant orientation relative to the sun. When they are oriented perpendicular to the sun then

$$g(r_j) = \delta(r_j - r) , \tag{14}$$

where r represents the zenith and azimuth angles of the solar beam.

Table 1.1. *Extinction coefficients for varous angle distribution functions. All except the heliotropic assume azimuthal symmetry. The beam zenith angle is θ, and the element inclination angle is θ_j. The parameter, x, for the ellipsoidal distribution, is the ratio of vertical to horizontal projections of canopy elements or $G(0)/G(\pi/2)$*

Horizontal inclination	
$G = \cos\theta$	$K = 1$
Vertical inclination	
$G = 2 \sin q/\pi$	$K = 2 \tan\theta/\pi$
Conical inclination, $\theta+\theta_j{\leqslant}\pi/2$,	
$G = \cos\theta \cos\theta_j,$	$K = \cos\theta_j$
Conical inclination, $\theta+\theta_j>\pi/2$	
$G = \cos\theta \cos\theta_j [1+2(\tan\beta - \beta)/\pi]$	$K = \cos\theta_j [1+2(\tan\beta - \beta)/\pi]$
$\cos\beta = 1/(\tan\theta \tan\theta_j)$	
Spherical (uniform) distribution	
$G = \frac{1}{2}$	$K = 1/(2\cos\theta)$
Heliotropic (leaves perpendicular to solar beam)	
$G = 1$	$K = 1/\cos\theta$
Ellipsoidal distribution	
$G = (x^2 \cos^2\theta + \sin^2\theta)^{1/2}/(A x)$	$K = (x^2 + \tan^2\theta)^{1/2}/(Ax)$

$A = A_1$ (eqn (12)) for $x >1$, $A = A_2$ (eqn (13)) for $x<1$, $A=2$ for $x=1$
A is closely approximated by $A = [x + 1.774 (x + 1.182)^{-0.733}]/x$

Fig. 1.3. The extinction coefficient, G, as a function of zenith angle for x values representing various canopy angle distributions.

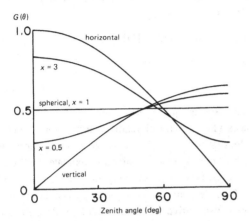

Fig. 1.4. The extinction coefficient, K, as a function of zenith angle for x values representing various canopy angle distributions.

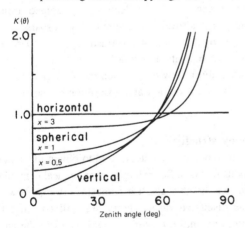

Using the extinction coefficients from Table 1.1, it is possible to determine the probability of a probe encountering 0, 1, 2,... canopy elements as it passes through a canopy. If canopy elements are randomly dispersed in space, then the number of contacts along a path through the canopy is a random variable which has a Poisson distribution function (Nilson, 1971). For a canopy which approximates the Poisson model, the probability of a probe traversing a distance through the canopy, s, in direction, r, without intersecting any canopy elements is

$$P_0(z,r) = \exp[-s\, u(z)\, G(\theta)] \ , \tag{15}$$

where $u(z)$ is the area density of canopy elements, and $G(\theta)$ is the appropriate extinction coefficient from Table 1.1. The probability of encountering n canopy elements in a transect of length s in direction r is

$$P_n(z,r) = [s\ u(z)\ G(\theta)]^n \exp[-s\ u(z)\ G(\theta)]/n! \ . \tag{16}$$

The mean number of intersections is

$$\mu(z,r) = s\ u(z)\ G(\theta) \ , \tag{17}$$

and the variance of the number of contacts is equal to the mean.

For a canopy which approximates a Poisson model, relative variance, or ratio of the variance to the mean is unity. Measurements of relative variance for a canopy are, therefore, useful for determining how closely real canopies approximate the Poisson model. When the relative variance for a canopy exceeds unity, the canopy is said to be clumped or underdispersed, and if it is less than unity, the canopy is overdispersed. Nilson (1971) discusses positive and negative binomial models to describe such canopies. Such models have been used by Monteith (1965) and others to describe the interaction of radiation with canopies.

The most obvious departure from a uniform dispersion of canopy elements occurs when leaves are clumped around a shoot, or when leaves are clumped around individual plants or rows of plants, as in a row crop, an orchard, or a sparse forest stand. Such situations were modelled by Allen (1974), Norman & Jarvis (1975) and Norman & Welles (1983) as clumps of vegetation, regularly or randomly dispersed, within which the foliage distribution did follow the Poisson model. Eqns (15) and (16) may therefore be used to model transmission and absorption of radiation by individual clumps.

Measurement of canopy structure

We now consider some methods which can be used to obtain the canopy parameters discussed in the previous sections. Recording of phenological data, plant populations, locations, and dimensions, heights, and leaf numbers is straightforward, though tedious, and will not be discussed further here. We will consider the determination of the area density function, $u(z)$, or its integral, $L(z)$ and the angle density function, $g(z,\theta_j)$, or its integrals, $G(\theta)$ or $K(\theta)$. We will also discuss some methods for determining the mean and variance of the number of intersections of a probe with canopy elements. Techniques for determining these parameters fall into three broad categories: direct measurement, indirect measurement, and allometric determination. Direct measurement methods are those where area and angle measurements are made directly on canopy elements to determine canopy parameters. Indirect methods require a model which relates some canopy response, such as light transmission or reflection, to the canopy structure parameters. The response is measured under appropriate conditions, and the model is inverted to determine the canopy structure parameters. Allometric methods are those which relate easily measured canopy properties to parameters which are more difficult to determine using empirically derived relationships. The best known of these are the relationships between sapwood area and leaf area in trees. We now briefly consider several direct methods and allometric methods, and then will treat indirect methods in some detail since they are not widely published elsewhere.

Direct measurement methods

Two methods have been suggested for direct determination of canopy properties. The stratified-clip method was introduced by Monsi & Saeki (1953), and has been widely used in studies of canopy structure. Briefly, it consists of defining a representative volume of foliage, usually using a wire frame, and dividing the volume into layers which can be clipped to determine area and leaf orientation. The sampling volume should be large enough to include a relatively large number of plants, (ground areas of 0.25–2 m^2 are typical). The canopy is then stratified according to height, leaf inclination angle, and leaf azimuth angle. A study such as that of Lemeur (1973) is typical, where up to six canopy layers were defined, with six elevation angle classes of 15 degrees each, and eight azimuth angle classes of 45 degrees each. Inclination and azimuth angles may be measured with a protractor and compass for each canopy element, and then that element is clipped and placed in a labelled polythene bag to await area measurement. The stratified-clip method is most useful in canopies such as grasses or small legumes, where plant population densities are high, and studies of individual plant characteristics would be difficult. This method is obviously not suitable for forests and other canopies where plants are widely spaced and large.

The second method is called the dispersed-individual-plant (DIP) method (Ross, 1981). The method consists of selecting 10–30 plants which are representative of the canopy and measuring characteristics of those plants, such as leaf area, spatial distribution of area, leaf angles, node heights, stem diameters, etc. Leaves are again cut from the plant as their locations and angles are determined. Samples are stratified at least according to height, inclination angle, and azimuth angle, and may also be stratified according to distance from the plant axis, to obtain some information on area density functions for individual plants. The DIP method is well suited to situations where plant densities are low and plants are large.

Several modifications have been suggested for decreasing the difficulty and improving the accuracy of these direct methods. Lang (1973) and Lang & Shell (1976) used a device consisting of three arms connected such that a pointer could be located at any point within a given canopy volume. Angles between the arms were measured using precision potentiometers and the voltages from the potentiometers were sensed using a computer-controlled data logger. The locations of points defining triangles on the surface of canopy elements are directly entered into the computer, and the areas, inclinations, and azimuths of the canopy elements are calculated. A number of portable logging devices with adequate computing capability for this application are now commercially available.

Allometric measurement methods

Allometric methods are discussed by Ross (1981), and some have been used extensively for obtaining the leaf area index of forest canopies (e.g. Grier & Waring, 1974; Kaufmann & Troendle, 1981; Marchand, 1983). Measurements show a linear relationship between sapwood area and leaf area for a given species and environment. Sapwood area is estimated, nondestructively, using increment borings. The primary difficulty of the method is the need to calibrate each species and to make adjustments for environment. Whitehead, Edwards & Jarvis (1984) suggested a refinement of the method which appears to eliminate some of the species variability. Water permeability of the sapwood was measured in addition to area. This accounted for the differences observed in the relationship between the two species studied. The height distribution of leaf area can be determined by measuring sapwood area at various heights.

Indirect measurement methods

There are many possible methods for indirect determination of canopy structure. Those to be discussed here all rely on the insertion of a physical or optical probe into the canopy to determine either the number of intersections of the probe with canopy elements within a given canopy volume, or the probability of traversing a given volume of canopy without intersecting any canopy element.

The first method is called the method of inclined point quadrats. It was first used in New Zealand by Levy & Madden (1933). Reeve, in an appendix to Warren–Wilson (1959) provided the necessary analysis for the method to be used to determine area density and mean inclination angle of canopy elements. Philip (1965) extended the analysis to determination of the inclination angle distribution function.

The second method is gap fraction analysis. Bonhomme *et al.* (1974) analysed fisheye photographs of canopies to determine the distribution of the gap fraction, and then used a simple method of mathematical inversion to estimate leaf area index. More sophisticated inversion methods can be used to determine both leaf area index and leaf angle distributions of canopies (e.g. Norman *et al.*, 1979).

The original inclined point quadrat method made use of a thin, pointed probe, which was inserted into the canopy at various zenith and azimuth angles. The location of contacts with canopy elements was recorded for a large number of insertions of the probe into a given canopy volume. Eqn (17) gives the relationship between the mean number of intersections of foliage elements, $\mu(z,r)$, and the area density, $u(z)$. For canopies with azimuthal symmetry, the relationship between area density and intersections with foliage elements is found by multiplying both sides of eqn (17) by $2\sin\theta\,d\theta$, integrating from 0 to $\pi/2$, and noting that

$$\int_0^{\pi/2} G(\theta)\sin\theta\,d\theta = 1 .$$

This gives

$$u(z) = \int_0^{\pi/2} (\mu(z,\theta)/s) \sin \theta \, d\theta . \tag{18}$$

The quantity $\mu(z,\theta)/s$ is the mean number of hits at height z in the canopy, averaged over all azimuth angles, per unit distance, s, traversed by the probe in direction θ. Once $u(z)$ is known, $G(\theta)$ may be obtained from eqn (17) by plotting $\mu(z,\theta)/[s\, u(z)]$ as a function of θ.

As previously noted, the relative variance, or ratio of variance of number of hits to mean number of hits, should be unity for a canopy which meets the assumption of uniform dispersion. The inclined point quadrat method could be used to estimate the variance needed for this calculation, though the number of probe insertions required to accurately determine the variance may be so large as to make the method impractical.

Several modifications to the inclined point quadrat method have been suggested. A laser probe was substituted for the pointed metal probe by Vanderbilt, Bauer & Silva (1979). The laser probe can be used for measurements on forest canopies where the use of the metal probe becomes practically impossible, but the data obtained are limited to just the location of the first intersection with the canopy. A second modification was suggested by Caldwell, Harris & Dzurec (1983) who used a motor-driven probe shaft with an encoder, and a fibre-optic device to record contacts with canopy elements, thus allowing automatic recording of the contacts.

The main disadvantage of the inclined point quadrat method is the labour required to obtain the data. Large numbers of probe insertions are needed to obtain reliable estimates of the canopy structure parameters. Also, the height of a canopy which can be analysed with the method is limited, unless the laser probe is used.

Gap fraction analyses for determining canopy properties are similar to inclined point quadrat analyses, except that only the probability of zero intersections (P_0 in eqn (15)) is known. Since the argument in the exponent of eqn (15) is just the mean number of intersections, the analyses presented here of gap fraction information could as well apply to inclined point quadrats.

Gap fraction measurements have generally been obtained using light as the probe. A measurement of the fractional length of a transect under a plant canopy that is exposed to direct beam radiation is sufficient for the calculation. Measurements can be made manually, by measuring the total length of sunflecks along a tape or metre stick, or with light sensors. In tall canopies, penumbral effects smudge the boundaries and only methods using sensors that differentiate between beam and diffuse are suitable. Lang, Xiang Yuequin & Norman (1985) describe a sensor that can be used to determine the gap fraction along a transect. Briefly, it consists of a filtered silicon cell at the end of a collimating tube. Filters limit the response of the silicon cell to wavelengths around 430 nm, and the collimator limits the view angle to 0.302 sr. A sighting tube aids in the alignment of the sensor so that it only sees beam radiation. The sensor is mounted on a track or cable so that it can be moved through the canopy at various heights. The gap fraction of the canopy is determined by measuring the mean transmission for beam radiation along the transect. Transects are made at several times on clear days with the sun at different elevation angles.

Fisheye photographs made either from below (Anderson, 1970; Bonhomme & Chartier, 1972; Fuchs & Stanhill, 1980) or from above the canopy (Bonhomme, Varlet Grancher & Chartier, 1974) can also be used for determining the gap fraction of canopies. A single fisheye photograph shows the gaps for all azimuth and zenith angles of the hemisphere seen by the lens, but measurements from several photographs are needed to give reliable estimates of gap fraction functions because of the variability of canopies. The photographs may be analysed, either manually (Anderson, 1970) or automatically (Bonhomme & Chartier, 1972) to determine the gap fraction. The photographs should be taken only under diffuse light (on an overcast day, or just before sunrise or after sunset). Beam radiation reflected from canopy elements, or the presence of the solar beam in the photograph can cause such non-uniform exposure that analysis of the photograph is difficult or impossible.

A modified form of eqn (15) is used to find the leaf area index and leaf angle distribution from the gap fraction data. If $G(\theta)$ is constant throughout the canopy, a canopy transmission coefficient, $\tau(z)$, for beam radiation can be defined as:

$$\tau(z) = \exp[-K(\theta,\theta_j)\, L(z)] . \tag{19}$$

Eqn (19) can be used in various ways to determine the leaf area index, and possibly also the leaf angle distribution function for a canopy. The simplest application is that of Bonhomme et al. (1974). Fig. 1.4 shows that the extinction coefficients for all angle distribution functions are near unity when the zenith angle is around 60 degrees. Bonhomme et al. used fisheye pictures, taken from above the canopy to determine the gap fraction at zenith angle of 57.5 degrees. They determined that the average K value for this zenith angle for all canopy types was 1.12 so leaf area index was calculated from eqn (19) using the measured gap fraction at 57.5 degrees and a K value of 1.12. Excellent agreement was observed between measured and predicted leaf area index up to L_0 of about 1.

Additional information about the canopy can be obtained from measurements of gap fractions at other zenith angles. If we consider the canopy to be composed of independent elements, each having a given area index, L_i, a given inclination angle, θ_i, and a given extinction coefficient, $K(\theta,\theta_i)$, then the probability of a ray penetrating the entire canopy is the product of the probabilities of a ray penetrating each of the populations or classes of individual canopy elements. Using eqn (19), we can therefore write

$$-\ln[\tau(\theta)] = \sum_i L_i\, K(\theta,\theta_i) . \tag{20}$$

If values for $\tau(\theta)$ are known for a number of values of θ, and $K(\theta,\theta_i)$ is computed for each zenith angle and leaf angle class then eqn (20) can be solved to give the L_i

values. If the number of angle classes, i, is equal to the number of zenith angles, eqn (20) represents a set of m linear equations in m unknowns, which is readily solved. If the number of zenith angles is greater than the number of angle classes, the problem is a linear least squares problem, and values of L_i are sought which minimise the sum of squares of differences between measured and predicted values of $\ln[\tau(\theta)]$. The sum of the values of L_i obtained in this way is the leaf area index, and the leaf angle distribution function can be estimated from the fraction of the leaf area in each leaf angle class. It is appropriate to minimise differences in $\ln \tau$ rather than τ because errors in the τ measurement are likely to be a fairly constant fraction of the measurement, and τ is therefore likely to have a log-normal distribution.

The values for L_i determined by standard least squares procedures are not constrained to physically realistic values, and can therefore be positive or negative. A negative leaf area index may give a good fit to the data, but is obviously not an acceptable solution. The least squares solution to eqn (20) must therefore be constrained to give only positive values for the L_i. Methods for constraining the solution are given by Menke (1984) and Lang *et al.* (1985). Menke gives FORTRAN code for implementing a least squares fit to data with positivity constraints.

A simpler method for finding area index assumes the ellipsoidal distribution and solves eqn (19) by a least squares method for x and L. We would like to find values for x and L which minimise

$$F = \Sigma \left(\ln \tau_i + K(\theta_i, x) L \right)^2 , \tag{21}$$

subject to the constraint, $x > 0$. The minimum can be found by solving $\partial F/\partial L = 0$ and $\partial F/\partial x$, simultaneously. From the first equation we obtain

$$L = -\Sigma \left[K(\theta_i, x) \ln \tau_i \right] / \Sigma K(\theta_i, x)^2 , \tag{22}$$

and from the second,

$$L = -\Sigma [\ln \tau_i \, \partial K(\theta_i, x)/\partial x] / \Sigma [K(\theta_i, x) \, \partial K(\theta_i, x)/\partial x] . \tag{23}$$

To solve the equations, L can be eliminated between them, and x found by the bisection method. Once x is known, L is found from eqn (22). Fig. 1.5 is a BASIC computer program which illustrates the bisection method and finds L and x from measured transmission coefficient data.

All of the inversion methods discussed so far rely on the assumption that the canopy elements are randomly dispersed in space. This is obviously not a good assumption for row crops before canopy closure, for coniferous trees or for canopies which never close, such as in desert vegetation. Canopies with heliotropic leaves or regular dispersion also violate this assumption. Lang *et al.* (1985) show examples of the errors that can arise when gap fraction inversion methods are applied to both row crops and crops with heliotropic leaves. In such cases a more sophisticated model and inversion method is needed. For row crops, the model of Mann *et al.* (1979) or

Norman & Welles (1983) could be used, along with the ellipsoidal angle distribution. For spaced plants, such as in orchards, the model of Norman and Welles could also be used. Inversion of these models follows principles similar to those already discussed, but requires more elaborate computer codes. Marquardt's (1963) method can be used to invert such models.

Fig. 1.5. BASIC program for finding leaf area index and x for a canopy from measurements of gap fraction at several zenith angles.

```
10 REM ******* ELLIPSOIDAL EXTINCTION COEFFICIENT
20 DEF FNK(Z,X)=SQR(X*X+Z*Z)/(X+1.774*(X+1.182)^-0.733)
25 REM ******* Z IS TAN(ZENITH ANGLE)
30 REM *******
40 PI=3.14159:DX=.01
50 INPUT "NUMBER OF ZENITH ANGLES";NZ
60 DIM Z(NZ),T(NZ)
70 FOR I=1 TO NZ
80    PRINT "ZENITH ANGLE";I;" - DEGREES";:INPUT Z(I)
90    PRINT "TRANSMISSION AT ";Z(I);"DEG";:INPUT T(I)
100   Z(I)=TAN(Z(I)*PI/180):T(I)=LOG(T(I))
110 NEXT
120 REM ******* FIND X USING BISECTION METHOD
130 XMAX=20:XMIN=0:X=1
140 S1=0:S2=0:S3=0:S4=0
150 FOR J=1 TO NZ:TZ=Z(J)
160    KB=FNK(TZ,X):DK=(FNK(TZ,X+DX)-KB)
170    S1=S1+KB*T(J):S2=S2+KB*KB:S3=S3+KB*DK:S4=S4+DK*T(J)
180 NEXT
190 F=S2*S4-S1*S3  :PRINT X,F
200 IF F<0 THEN XMIN=X ELSE XMAX=X
210 X=(XMAX+XMIN)/2
220 IF (XMAX-XMIN)>.01 THEN GOTO 140
230 REM ******* FIND LAI AND PRINT RESULTS
240 L=-S1/S2:PRINT "LEAF AREA INDEX=";L
250 PRINT "RATIO OF VERTICAL TO HORIZONTAL PROJECTIONS=";X
260 PRINT "MEAN LEAF ANGLE - DEGREES=";90*(0.1+0.9*EXP(-0.5*X))
270 PRINT
280 PRINT "ZENITH ANG.","MEASURED T","PREDICTED T"
290 FOR J=1 TO NZ
300    PRINT ATN(Z(J))*180/PI,EXP(T(J)),EXP(-FNK(Z(J),X)*L)
310 NEXT
```

Future directions

The application of inversion methods to measurements of canopy structure is obviously in its infancy. Within the next few years, new methods are likely to be found which will dramatically decrease the effort required to obtain canopy structure information and will increase its accuracy. In addition to those methods already discussed, others using reflected or transmitted radiation, or thermal radiation could be used. The use of computed tomography methods (Vanderbilt, 1985) (which are, of course, inversion methods) is likely to increase in the future.

Continued development of inverse methods will require improved models of canopy structure and the interaction of radiation with canopies. Additional work is needed to model canopies with row structure and with regular and random individual plant spacing. Additional emphasis needs to be given to the relationship between canopy characteristics and the transmission or reflection of radiation. Models of spatial variation of canopy radiation may provide useful information about canopy properties. The work of Norman, Miller & Tanner (1971) and Norman & Jarvis (1975) on modelling gap size distributions of canopies should be extended and evaluated as a possible inverse method. Geostatistical methods (Burgess & Webster, 1980) might be useful for modelling spatial variation of canopy properties.

New probes should also be developed. Optical (Vanderbilt, 1985) as well as acoustic (Shibayama & Akiyama, 1985) methods show some promise, and probes using electromagnetic radiation outside the visible wavebands may have application.

References

Allen, J.H. Jnr (1974). Model of light penetration into a wide-row crop. *Agronomy Journal*, **66**, 41–7.

Anderson, M.C. (1966). Stand structure and light penetration. II. A theoretical analysis. *Journal of Applied Ecology*, **3**, 41–54.

Anderson, M.C. (1970). Radiation climate, crop architecture and photosynthesis. In *Prediction and Measurement of Photosynthetic Productivity*, ed. J. Šĕtlik, pp. 71–8. Wageningen: PUDOC.

Bonhomme, R. & Chartier, P. (1972). The interpretation and automatic measurement of hemispherical photographs to obtain sunlit foliage area and gap frequency. *Israel Journal of Agricultural Research*, **22**, 53–61.

Bonhomme, R., Varlet-Grancher, C. & Chartier, P. (1974). The use of hemispherical photographs for determining the leaf area index of young crops. *Photosynthetica*, **8**, 299–301.

Burgess, T.M. & Webster, R. (1980). Optimal interpolation and isarithmic mapping of soil properties. I. The semi-variogram and punctual kriging. *Journal of Soil Science*, **31**, 315–331.

Caldwell, M.M., Harris, G.W. & Dzurec, R.S. (1983). A fiber optic point quadrat system for improved accuracy in vegetation sampling. *Oecologia*, **59**, 417–18.

Campbell, G.S. (1986). Extinction coefficients for radiation in plant canopies calculated using an ellipsoidal inclination angle distribution. *Agricultural and Forest Meteorology*, **36**, 317–21.

Charles-Edwards, D.A.& Thornley, J.H.M. (1973). Light interception by an isolated plant, a simple model. *Annals of Botany*, **37**, 919–28.

Fuchs, M. & Stanhill, G. (1980). Row structure and foliage geometry as determinants of the interception of light rays in a sorghum row canopy. *Plant, Cell and Environment*, **3**, 175–82.

Grier, C.C. & Waring, R.H. (1974). Conifer mass related to sapwood area. *Forest Science*, **20**, 205–6.

Kaufmann, M.R. & Troendle, C.A. (1981). The relationship of leaf area and foliage biomass to sapwood conducting area in four subalpine forest tree species. *Forest Science*, **27**, 477–82.

Lang, A.R.G. (1973). Leaf orientation of a cotton plant. *Agricultural Meteorology*, **11**, 27–51.

Lang, A.R.G. & Shell, G.S.G. (1976). Sunlit areas and angular distribution of sunflower leaves for plants in single and multiple rows. *Agricultural Meteorology*, **16**, 5–15.

Lang, A.R.G., Xiang Yueqin & Norman, J.M. (1985). Crop structures and the penetration of direct sunlight. *Agricultural and Forest Meteorology*, **35**, 83–101.

Lemeur, R. (1973). A method for simulating the direct solar radiation regime in sunflower, Jerusalem artichoke, corn and soybean canopies using actual stand structure data. *Agricultural Meteorology*, **12**, 229–47.

Levy, E.B. & Madden, E.A. (1933). The point method of pasture analysis. *New Zealand Journal of Agriculture*, **46**, 267–79.

Mann, J.E., Curry, G.L. & Sharpe, P.H. (1979). Light interception by isolated plants. *Agricultural Meteorology*, **20**, 205–14.

Marchand, P.J. (1983). Sapwood area as an estimator of foliage biomass and projected leaf area for *Abies balsamea* and *Picea rubens*. *Canadian Journal of Forest Research*, **14**, 85–7.

Marquardt, D.W. (1963). An algorithm for least-squares estimation of nonlinear parameters. *Journal of the Society for Industrial and Applied Mathematics*, 11, 431–44.

Menke, W. (1984). *Geophysical Data Analysis: Discrete Inverse Theory*. New York: Academic Press.

Monsi, M. & Saeki, T. (1953). Über den lichtfaktor in den Pflanzengesellschaften und seine Bedeutung für die Stoffproduktion. *Japanese Journal of Botany*, 14, 22–52.

Monteith, J.L. (1965). Light distribution and photosynthesis in field crops. *Annals of Botany*, 29, 17–37.

Monteith, J.L. & Unsworth, M.H. (1990). *Principles of Environmental Physics*, second edition. London: Edward Arnold.

Monteith, J.L. (1985). Measurement – a game of snakes and ladders. In *Instrumentation for Environmental Physiology*, eds. B. Marshall & F.I. Woodward, pp. 1–4. Cambridge University Press.

Nilson, T. (1971). A theoretical analysis of frequency of gaps in plant stands. *Agricultural Meteorology*, 8, 25–38.

Norman, J.M. (1979). Modelling the complete crop canopy. In *Modification of the Aerial Environment of Plants*, eds. B.J. Bardield & J.F. Gerber, pp. 249–77. St Joseph, Michigan: ASAE.

Norman, J.M. & Jarvis, P.G. (1975). Photosynthesis in Sitka spruce (*Picea sitchensis* (Bong.) Carr.). V. Radiation penetration theory and a test case. *Journal of Applied Ecology*, 12, 839–78.

Norman, J.M. & Welles, J.M. (1983). Radiative transfer in an array of canopies. *Agronomy Journal*, 75, 481–8.

Norman, J.M., Miller, E.E. & Tanner, C.B. (1971). Light intensity and sunfleck-size distributions in plant canopies. *Agronomy Journal*, 63, 743–8.

Norman, J.M., Perry, S.G., Fraser, A.B.& Mach, W. (1979). *Remote sensing of canopy structures*. Proceedings of the 14th Conference of Agricultural and Forest Meteorology, pp.184–5. Boston: American Meteorological Society.

Pereira, A.R. & Shaw, R.H. (1980). A numerical experiment on the mean wind structure inside canopies of vegetation. *Agricultural Meteorology*, 22, 303–18.

Philip, J.R. (1965). The distribution of foliage density with foliage angle estimated from inclined point quadrat observations. *Australian Journal of Botany*, 13, 357–66.

Ross, J. (1981). *The Radiation Regime and Architecture of Plant Stands*. The Hague: W. Junk.

Ross, J. & Nilson, T. (1965). Propuskanye pryamoi radiatsii solntsa sel'skokozyaistvennymi poserami. *Tartu, Issledovaniya po Fizike Atmosfery, Akademiya Nauk Estonskoi SSR*, 6, 25–64.

Shell, G.S.G. & Lang, A.R.G. (1975). Description of leaf orientation and heliotropic response of sunflower using directional statistics. *Agricultural Meteorology*, 15, 33–48.

Shibayama, M. & Akiyama, T. (1985). A portable field ultrasonic sensor for crop canopy characterization. *Remote Sensing of Environment*, 18, 269–79.

Vanderbilt, V.C. (1985). Measuring plant canopy structure. *Remote Sensing of Environment*, 18, 281–94.

Vanderbilt, V.C., Bauer, M.E. & Silva, L.F. (1979). Prediction of solar irradiance distribution in a wheat canopy using a laser technique. *Agricultural Meteorology*, 20, 147–60.

Warren-Wilson, J. (1959). Analysis of the spatial distribution of foliage by two-dimensional point quadrats. *New Phytologist*, 58, 92–101.

Warren-Wilson, J. (1965). Stand structure and light penetration. I. Analysis by point quadrats. *Journal of Applied Ecology*, 2, 383–90.

Warren-Wilson, J. (1967). Stand structure and light penetration. III. Sunlit foliage area. *Journal of Applied Ecology*, **4**, 159–65.

Whitehead, D.W., Edwards, R.N. & Jarvis, P.B. (1984). Conducting sapwood area, foliage area, and permeability in mature trees of *Picea sitchensis* and *Pinus contorta*. *Canadian Journal of Forest Research*, **14**, 940–7.

G. RUSSELL, P.G. JARVIS AND
J.L. MONTEITH

2. Absorption of radiation by canopies and stand growth

Growth analysis – old and new

When a canopy of leaves is sunlit, photosynthesis proceeds at a rate which depends on how photons are distributed over individual elements of the foliage and on the relationship between photosynthetic rate and irradiance for each foliage element. In principle, therefore, photosynthesis by a canopy, expressed per unit of ground area rather than per unit leaf area, can be estimated from a statistical description of irradiance as a function of leaf disposition. In many models of productivity, this is a central and complex component. In practice, however, modelling can often be greatly simplified with little sacrifice of precision by exploiting the observation that, at least during vegetative growth, uniform stands produce dry matter at a rate which is almost proportional to the amount of radiant energy intercepted by the canopy. In this chapter we consider the theoretical basis of this relationship, its experimental verification, and its usefulness for exploring the dependence of growth on environmental variables in general.

Traditional growth analysis is based on the observation that, when *single* plants are exposed to a more or less constant environment, their rate of growth is approximately proportional to their weight and to their leaf area until a significant fraction of older foliage is shaded by younger foliage. Consequently, relative growth rate (RGR) and net assimilation rate (NAR) are conservative indices of growth initially. Extension of this procedure from single plants to stands of arable crops was made possible by the introduction of leaf area index (L) as a measure of the foliage in a canopy – a development by Watson (1947) which had far-reaching implications for crop ecology. However, mutual shading of foliage in most crop stands begins much earlier in the life of a plant than when it grows in isolation. The RGR and NAR of field crops, therefore, decrease as L increases, always confusing and often obscuring the dependence of growth rate on environmental variables.

One way out of this difficulty emerges clearly from the recognition that growth rate is often proportional to the amount of radiation intercepted by the canopy. If, for the moment, the transmitted fraction is assumed to obey Beer's law with L as a measure of optical depth in the canopy (Kasanga & Monsi, 1954), the increase of dry weight, dW, in a time interval dt, can be written as:

$$dW/dt = C[1 - \exp(-KL)] \tag{1}$$

where C is the maximum possible growth rate and K is an attenuation coefficient with a value often between 0.4 and 0.6. Expanding the exponential term and dividing by L gives NAR as:

$$(dW/dt)/L = CK(1 - KL/2 + ...) , \tag{2}$$

where higher-order terms are omitted.

The observation that NAR decreases almost linearly with increasing L (Watson, 1958) is, therefore, entirely compatible with a linear relation between growth and radiation. This relationship provides the basis for a new form of growth analysis in terms of:

(i) the fraction, i, of incident radiation intercepted by a canopy; this depends on its *structure* and we shall call it the radiation interceptance;

(ii) the amount of dry matter produced per unit of radiation intercepted by a canopy; this depends on its *function* and we shall call it the 'dry matter:radiation quotient' (ε). (This quantity is often described as an 'efficiency' but it is not a ratio with a maximum value of 1.)

Thus biomass (W) can be considered as the time-integrated product of three factors

$$W = \int \varepsilon i Q \, dt , \tag{3}$$

where Q is the daily radiation incident on the top of the canopy and t is measured in days (Fig. 2.1). The proportion of radiation *absorbed* by the canopy can be used instead of the proportion intercepted, in which case the value of ε would have to be adjusted to allow for reflection. If ε is essentially constant it can be removed outside the integral sign. In some environments Q can also be considered constant over the growing season and the equation reduces to the product of mean interceptance, daily radiation, dry matter:radiation quotient and time.

Interpretation of intercepted radiation

The ratio of the average radiant flux density below a canopy to the corresponding value above is the fractional transmission or canopy transmittance, τ, and the canopy interceptance, i, is often estimated as $(1 - \tau)$. To be more precise, $(1 - \tau)$ is the sum of the absorbed and reflected fractions (Hipps, Asrar & Kanemasu, 1983), the former providing energy for canopy processes and the latter representing a loss of energy from the system (Fig. 2.2).

Because it is primarily the fractional absorption of the quantum flux that determines photosynthetic rate and, therefore, the production of assimilate, it is useful to derive a relation between this quantity and the more commonly measured interception of total solar radiation. Goudriaan (1977) showed that the application of the Kubelka–Munk equations to radiative transfer in uniform canopies provided a good estimate of the relationship between L and radiation transmission provided that:

(1) the sun was more than 20° above the horizon,
(2) the spatial distribution of leaves was random and
(3) the distribution of leaf angles was spherical. Taking his equations and neglecting second-order terms, the fractional transmission, τ, in a specified waveband is given by:

$$\tau = \exp\left(-KL\right) , \tag{4}$$

and the fractional reflection by:

$$\rho = \rho' - (\rho' - \rho_s)\tau^2 , \tag{5}$$

where ρ_s is the reflectivity of the soil and ρ' is the reflectivity of a stand with such a large leaf area that τ is effectively zero. The theory shows that ρ' can be calculated as $(1 - \alpha^{1/2})/(1 + \alpha^{1/2})$ where α is the mean absorption coefficient of leaves for a particular waveband. The attenuation coefficient K is $\alpha^{1/2} K_b$, where K_b is the coefficient for a stand with the same geometry but with completely black leaves.

Fig. 2.1. A diagram to show the empirical relationship between biomass and intercepted radiation. ε is the slope of the line CAB. Taking the biomass corresponding to point A, a change in ε will lead to a change in biomass along the line YAX: a change in the interception of radiation will lead to a change in biomass along BAC. W_m is the biomass which would result if all of the incident radiation were to be intercepted. Q_{im} is the radiation incident on the top of the canopy.

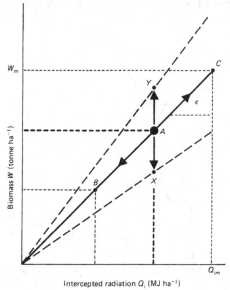

Measurements of radiation within the canopy are often made with instruments sensitive to the whole solar spectrum whereas it may be the profile of photosynthetically active radiation (PAR, 400–700 nm) that is required. The ratio of attenuation coefficients for PAR (K_p) and for total solar radiation (K_T), which can be calculated from the above equations, is therefore of interest:

$$x = K_p/K_T = (\alpha_p/\alpha_T)^{1/2} ,$$ (6)

and measurements of transmitted radiation in the two wavebands usually give values of x between 1.3 and 1.4 (Green, 1984). It follows from eqns (4) and (6) that:

$$\tau_p = \tau_T^{(\alpha_p/\alpha_T)^{1/2}} .$$ (7)

Because the quantum content of PAR is almost constant within canopies at a value of 4.6 mol MJ^{-1} (McCartney, 1978), the value of K for the transmission of quanta is effectively the same as for the transmission of PAR.

Allowing for the fraction of PAR reflected by the soil ($\rho_s\tau_p$), the fraction absorbed by the canopy is:

$$\alpha_p = 1 - \rho_p - \tau_p + \rho_s\tau_p ,$$

$$= 1 - (1 - \rho_s)\tau_p - \rho_p' + (\rho_p' - \rho_s)\tau_p^2 .$$ (8)

Fig. 2.2. A diagram to demonstrate what is meant by intercepted (Q_i) and absorbed (Q_a) radiation. Q is the incident radiation and ρQ the radiation reflected upwards from the vegetation. $Q_\tau (=\tau Q)$ is the radiation transmitted through the overstorey and $\rho_s Q_\tau$ is that part of the transmitted radiation that is reflected upwards by the soil and ground flora.

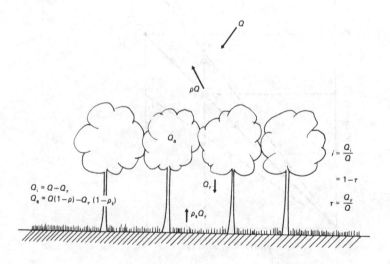

$$Q_i = Q - Q_\tau$$
$$Q_a = Q(1-\rho) - Q_\tau (1-\rho_s)$$

$$i = \frac{Q_i}{Q}$$
$$= 1 - \tau$$
$$\tau = \frac{Q_\tau}{Q}$$

Taking representative values of $\rho_s = 0.1$ and $x = 1.35$, it can be shown from eqns (7) and (8) that the difference between the fractional absorption of PAR and the fractional interception of total radiation is usually less than 0.1 (see Fig. 2.3). The difference between the fraction of PAR intercepted and the fraction absorbed is simply the fraction reflected by the canopy, about 3%, and the reflected fraction of the quantum flux is similar.

Over a whole growing season τ_T decreases from 1.00 to about 0.05 in a well-managed arable field crop and the mean value of the intercepted fraction $(1 - \tau_T)$ underestimates α_p by about 10%.

When W, $i(t)$ and Q are measured for stands of the same species growing in the same environment but managed differently, eqn (3) may be used to derive values of ε and $\int i(t)dt$, and these can be related to management practices. Caution is needed, however, when comparing the productivity of stands grown at different levels of mean irradiance. Because the relationship between photosynthetic rate and quantum flux density is not strictly linear, ε cannot be completely independent of the degree of variation in irradiance. But because many leaves in a canopy are exposed to relatively weak light for most of the day, ε is usually a weak function of irradiance for species with the C_3 pathway of CO_2 fixation growing in temperate cloudy climates and for C_4 species in all climates.

If we are interested in harvesting only part of a plant, for example stem, grain or oil content, the analysis can be extended by assuming that the harvested material is a constant proportion of the biomass at maturity. However, more complex relationships are needed when harvest index changes significantly with treatment.

This simple model can be readily extended into a simulation of growth by making ε and i functions of time, environmental variables and treatments, using solar radiation as the driving variable and incrementing biomass at convenient intervals. To complete

Fig. 2.3. Fractional absorption of PAR, α_p, (full lines) and fractional interception of total radiation, $(1 - \tau)$ (dashed lines), as functions of leaf area index for three values of the attenuation coefficient for black leaves K_b, $(x = 1.34)$.

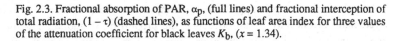
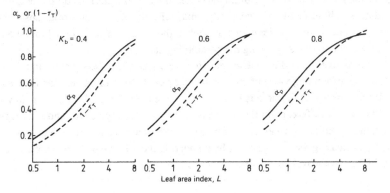

the model, all that is then required is a development sub-routine to control the duration of different phases of growth.

Measurement of intercepted radiation

One of the most common methods of measuring the interception of radiation by a stand is to install, above and below the canopy, tube solarimeters which integrate radiation over an area of about 20×900 mm (Szeicz, Monteith & dos Santos, 1964). Another method, more suitable in stands of larger scale, such as maize or trees, is to traverse a sensor through the stand along a track below the canopy over a transect of many metres in length (e.g. Norman & Jarvis, 1974).

Most tube solarimeters respond to radiant energy over the whole solar spectrum, but it is possible to pair instruments with and without a filter that transmits infra-red radiation only and to obtain the interception of photosynthetically useful radiant energy by difference (Szeicz et al., 1964; Palmer, 1980). Alternatively, the flux of photosynthetically useful quanta may be measured either with a line sensor with a quantum response rather than an energy response, or by traversing a small quantum sensor below the canopy. A distributed array of, say, 5–10 point sensors below the canopy may also give an adequate measure of interception, if the radiation or quantum flux is being totalled over several days, since the daily course of the sun considerably reduces the instantaneous, point to point variation.

Ideally, intercepted radiation should be measured directly using tube solarimeters, or quantum sensors, that are installed for the life of the crop. However, variation in transmitted radiation and the consequent requirement of replication makes this approach expensive for large-scale field trials with a number of treatments. It is possible, however, to move sensors around to sample separate plots one after the other (Gallo & Daughtry, 1986), provided that the measurements are made in the four-hour period centred on solar noon when irradiance is strongest. Russell & Ellis (1988) have used this approach to sample 64 plots at weekly intervals. Systematic errors may be introduced if the proportion of radiation absorbed is significantly affected by the diurnal variation in solar elevation or by the degree of cloudiness. These errors are generally small and the approach works best when treatments at one site are compared and when the receipt of intercepted radiation is calculated over a period spanning many sampling dates.

An alternative method is to determine foliage cover from photographs and then to find the corresponding value of interceptance by calibration or calculation (Steven et al., 1986). One difficulty here is that at sites where the solar angle at noon is low we would expect a difference in the relationship between foliage cover and intercepted radiation depending on whether the day is overcast or not. There are also methods for directly estimating the foliage cover by comparing the relative proportion of reflected light in different wavebands (Birnie et al., 1987).

If the radiation transmitted through the canopy follows eqn (1) and the attenuation coefficient is known, the intercepted fraction can be calculated from the leaf area index. This may be found either directly by measurement of the leaf area of a sample or indirectly with a vertical (Russell *et al.*, 1982) or inclined (Warren–Wilson, 1963) point quadrat, using either a sharply pointed probe or the beam of the sun (Lang, Yuequin & Norman, 1985; Lang & Yuequin, 1986; Lang, 1986). The appropriate method depends on the particular circumstances, including the purpose of the work, the equipment available and the height of the canopy.

The number and distribution of sample areas for measurement needed to obtain statistically reliable results depends on the depth and form of the canopy, and especially on whether the component plants are distributed randomly, regularly, as in row, plantation or orchard crops, or in clumps. Methods have been devised to allow the computation of interceptance for discontinuous plant stands such as orchards (Jackson & Palmer, 1981), vineyards (Smart, 1973) and widely spaced row crops (Shuttleworth & Wallace, 1985).

The value of ε, the dry matter:radiation quotient, is often calculated as the slope of the best-fit line on a plot of cumulative dry matter against cumulative intercepted or absorbed PAR, points being obtained by sequential sampling throughout the growing season (e.g. Cannell *et al* ., 1987). This is not statistically justifiable, since the points are not independent, each being based on the value of the preceding one. To avoid this problem, the finite *rate* of increase in dry matter, $\Delta W/\Delta t$, should be plotted against the finite *rate* of absorption of PAR, $\Delta Q_i/\Delta t$, using pairs of points with values obtained independently. Alternatively, the rate of increase in dry matter can be obtained as the slope of the best-fit line on the plot of weight against time using the methods of time-series analysis, provided that there are enough points and that observations are made at equal intervals.

Maximum value of the dry-matter:radiation quotient and annual productivity

If we take, as a practical working value, a quantum requirement for photo-synthesis by leaves in daylight of 20 mole quanta per mole of net CO_2 assimilated for C_3 plants (Björkman, 1981) and combine this with the equivalence of 4.6 mole quanta per MJ of PAR (or 2.3 mole quanta MJ^{-1} (total solar)), then 2.3/20 mole CO_2 are assimilated per MJ of total solar radiation absorbed. If we additionally assume an equivalence of 1 mole CO_2 to 1 mole of dry matter of 'typical composition', we obtain a conversion factor for mass of CO_2 to mass of dry matter of approximately 0.50 (Penning de Vries, Brunsting & van Laar, 1974; Ledig, Drew & Clark, 1976). Then we could expect a potential dry matter:radiation quotient of $2.3/20 \times 44 \times 0.5 = 2.5$ g MJ^{-1} (total) or approximately 5.0 g MJ^{-1} (PAR). This estimate is about 1.4–2.0 times larger than many estimates which fall in the range 1.2– 1.7 g MJ^{-1} (total) for annual crops (Table 2.1). The lower values of ε realised in practice result from radiation saturation, environmental stresses, such as low temperatures and shortage of water, respiratory losses and losses associated with the mortality of leaves and roots and the costs of their replacement.

In temperate climatic regions the annual input of solar radiation ranges from 2.3 to 6.0 GJ m^{-2}, depending on latitude, daylength and cloudiness. For a climate with an average annual solar radiation receipt of 4.0 GJ m^{-2}, the potential annual dry matter production for vegetation completely intercepting all the solar radiation throughout the year is therefore about 100 tonnes ha^{-1}. This is about 2.5 times the highest productivities recorded for both herbaceous annual vegetation and perennial tree stands (Jarvis, 1981). This shortfall occurs because of the lower values of ε achieved in practice, as discussed above, and because the seasonal cycle of leaf phenology does not allow complete radiation interception every day of the year. In many instances, canopy interceptance is quite small for much of the year.

Factors affecting the dry matter:radiation quotient, ε

The value of ε has been estimated for a large number of species, sites and treatments (see Gosse *et al.*, 1986; Cannell *et al.*, 1987, for recent compilations). A representative selection is given in Table 2.1. However, it is often difficult to compare estimates from different experiments because of disparities in measurement technique and in the nature of the observations. Apart from differences in the specification of the measure of radiation used, e.g. total solar, PAR or quantum flux, and failure to distinguish between intercepted and absorbed radiation, there are three main reasons for discrepancies. First, the whole plant is not always measured either because only the tops were sampled or because loss of plant parts by death (turnover) or grazing was significant. For example, in one study on young *Pinus sylvestris* in Sweden it was found that up to 75% of the annual photosynthetic production was

Table 2.1. *Estimates of the dry matter:radiation quotient (g mJ^{-1}) for various crops amply supplied with nutrients and water. The figures are based on intercepted solar radiation and have been determined for the main period of growth. Root weights have been included except where marked with an asterisk*

Crop	ε	Reference
Brassica napus	1.1	Mendham, Shipway & Scott (1981)
Cajanus cajan	1.2	Hughes, Keatinge & Scott (1981)
Eucalyptus globulus	0.5*	Linder (1985)
Pinus radiata	0.9*	Linder (1985)
Pisum sativum	1.5	Heath & Hebblethwaite (1985)
Salix viminalis	1.4*	Cannell *et al.* (1987)
Solanum tuberosum	1.4	Scott & Allen (1978)

partitioned to fine roots, which subsequently died (see Jarvis & Leverenz, 1983). The second is the interception of radiation by non–photosynthetic parts of the plants, for example by tree trunks or by flowers. In oilseed rape (*Brassica napus*) up to 60% of incoming radiation can be reflected or absorbed by the mass of yellow flowers at the top of the canopy for a short period (Mendham, Shipway & Scott, 1981). The final reason is related to the length of the measurement period and will be considered later. For local comparisons between treatments these problems do not matter but other comparisons should be made with caution.

During vegetative growth, many agricultural crops (Monteith & Elston, 1983) and some other plant stands have remarkably similar dry matter:radiation quotients. A typical figure for C_3 plants is 1.4 g of dry matter for each MJ of total solar radiation intercepted (Monteith, 1977). C_4 species, for example maize (*Zea mays*) (Bonhomme *et al.*, 1982), seem to exhibit somewhat larger values of ε. Nevertheless, there is still uncertainty as to the circumstances in which there are differences either between taxa (family, species, genotype) or over the period of growth. Two features of plant response may be invoked to explain any differences encountered. First the relationship between accumulated photosynthate and dry weight is not necessarily constant. Although there is a large potential for such variation, in practice it is small except in a few, well-defined situations. For example, variations in the ash content of plant material can generally be ignored since the ash content is normally less than 8% of the dry weight. On the other hand, the energy content of the different components of plant material (monosaccharide, polysaccharide, lipid and protein) varies widely, with the result that changes in biochemical composition may also cause the energy content of the dry matter to vary. Differences in the composition and energy content of dry matter between types of plants (e.g. herbs and trees) and species have been recorded (Penning de Vries *et al.*, 1974; Sinha, Bhargava & Goel, 1982) but they are likely to be significant only in crops like oil palm (*Elaeis guineensis*), and then only during the phase of development when oil is being synthesised (Squire, 1986). If we assume a typical dry matter composition with an energy content of *ca* 20 kJ g^{-1}, a dimensionless energy conversion efficiency is simply obtained as $2.\varepsilon\%$. Thus, if ε is 1.5 g MJ^{-1}, the efficiency of conversion of solar radiation to biomass is 3.0%.

Assimilation may also be lost by respiration to support non-photosynthesising tissues, like the woody stems of many perennials, or directly to support symbionts, like mycorrhizas or bacteria. Few values of ε are as yet available for woody perennials or for trees (Jarvis & Leverenz, 1983). Making use of eqn (4), with assumed values of K to estimate the canopy transmittance and, hence, the intercepted radiation, Linder (1985) calculated an ε on the basis of above ground biomass of 0.9 g MJ^{-1} (PAR) (i.e. 0.45 g MJ^{-1} of total solar) for *Eucalyptus globulus* in Australia over its first ten years and an average value of 1.7 g MJ^{-1} (PAR) (i.e. 0.85 g MJ^{-1} of total solar) for a number of evergreen stands, up to 55 years of age, including the conifers *Pinus radiata* in Australia, *Pinus sylvestris* in Sweden and England and *Pinus nigra* and

Picea sitchensis in Scotland. These two values are well below the typical values for annual, herbaceous crops referred to earlier. On the other hand Cannell *et al.* (1987) obtained comparable values of 0.99 and 1.38 g MJ^{-1} (total solar) for short-rotation willows (*Salix viminalis*, clone Bowles hybrid) in Scotland in successive years without and with irrigation respectively. This lower ε seems to be associated with the accumulation of woody biomass. The generally low values of ε achieved by leguminous species (Gosse *et al.*, 1986) may be partly attributable to the demands of the nitrogen-fixing rhizobia in the root nodules.

Naturally, any variable that affects the rate of photosynthesis significantly will also affect the dry matter:radiation quotient. PAR is one such variable. A linear response of growth to absorbed PAR is to be expected as long as the canopy is not exposed to saturating irradiances for significant parts of the growing season. The irradiance at which the rate of photosynthesis is effectively saturated will depend on both leaf properties and canopy structure, being less for planophile canopies than for erectophile canopies or for canopies made up of small, grouped leaves, such as occur on heaths and conifers. Whilst the photosynthetic rate of individual leaves in C_3 species may reach a maximum at quite small values of solar irradiance, say 250 W m^{-2} or a quantum flux density of 500 μmol m^{-2} s^{-1} in C_3 species, the saturating irradiance is much higher for groups of leaves around shoots, and many vegetation canopies may never even approach saturation (Jarvis & Leverenz, 1983). The crucial question is how much of the absorbed radiation is received at irradiances above the saturation irradiance for the canopy. The answer depends on both the structure of the canopy and the radiation climate of the location being considered. At one site in the south of France (latitude 44 $^\circ$N) in July about 30% of PAR was received at irradiances exceeding 250 W m^{-2} PAR (Bedel, 1983). However, this is an extreme case, since the daily PAR receipts in July were 15 MJ m^2, and the proportion above 250 W m^2 would have been much less for the growing season as a whole. At higher latitudes in summer, radiation receipts are spread over a longer day so that the proportion of PAR above saturation is correspondingly less. In most field situations, the effect of the shape of the photosynthesis:PAR response curve of individual leaves on ε for the canopy is likely to be small. However, Kasim & Dennett (1986) have shown that field beans (*Vicia faba*) exhibit higher values of ε when grown in shade, although they suggest that the shape of the photosynthesis:PAR response curve is not the only factor involved. In some species, the leaves in the canopy show diurnal movements which have the result of altering the canopy saturation irradiance (Ehleringer, in this volume).

The value of ε may change over the life of the plant, possibly as a consequence of ontogenetic changes in canopy structure as well as possible effects of alterations in sink activity on photosynthetic rate. It is well known that photosynthetic rate per unit leaf area changes with leaf age and with ontogeny (Šesták, 1981). Nonetheless, in species where leaves are produced continuously, the average leaf age may not change much, the old leaves dying regularly and being replaced by newly formed ones. However, once the last leaf has appeared the average leaf age increases with time and consequently ε is commonly larger when assessed over the vegetative phase of annual plants alone than when it is assessed for the whole period up to maturity (see, for example, Gallagher & Biscoe, 1978). Harper (in this volume) has pointed out that leaf area index is not a very informative way to describe a canopy since a constant L may be obtained either from a few long-lived leaves or from many short-lived ones, even within the same species. Defoliation can rejuvenate a canopy: for example, in a sward of grass, growth rate declines, even when radiation interception is complete, unless the sward is grazed to encourage the production of new leaves. In long-lived stands of trees, ε declines with age as the ratio of respiring surface area to assimilating leaf area increases, so that in old-growth stands there may be no increase in dry matter from year to year, despite substantial radiation interception and photosynthesis (see Jarvis & Leverenz, 1983).

Plant breeders are interested in whether there are any differences in ε between genotypes of the same species. The evidence is conflicting. Hughes *et al.* (1987) found differences between erect-leaved and prostrate-leaved types of chickpea, and Bonhomme *et al* . (1982) found differences between three maize genotypes, although in the latter case the standard errors were relatively large. On the other hand, Heath & Hebblethwaite (1985) found differences in ε between three pea (*Pisum sativum*) varieties with contrasting leaf phenotype only under conditions of water shortage. In experiments over three years with four different varieties of potatoes, ε varied between 1.43 and 1.94 g MJ^{-1} (solar) (D.K.L. McKerron, pers. comm.). On the whole the differences between these genotypes were consistent but the values of ε varied somewhat from year to year for no obvious reasons and, in a year without irrigation, differences between the varieties changed. Thus, within a species, ε may depend either on the interactions between genotype and environment as they affect processes such as photosynthesis and transpiration or on canopy structure as it affects the amount of radiation penetrating the canopy and reaching the lower leaves. In the latter case ε may be manipulated by management decision. Higher productivity of erectophile cereals, for example, results when they are planted close together with a larger number of plants per unit area.

Factors affecting the proportion of PAR absorbed

In many crops, it is absorbed radiation rather than ε which is most closely correlated with biomass. The absorbed radiation depends not only on the daily PAR receipts but also on the duration of the various phases of growth. The proportion of incident PAR absorbed depends primarily on canopy structure, although the radiation climate may also be important. In cloudy weather, for example, radiation is received from all angles and as a result penetrates further into the canopy. Near the equator, most beam radiation comes from zenith angles near the vertical, while at higher latitudes in cloudless conditions the predominant source of radiation is at progressively larger zenith angles. These considerations may not, however, be treated as independent of canopy structure, since the growth form of the plants and the orientation of the leaves may be conditioned by the angular distribution of the incident radiation.

The structural property of a canopy that has the largest effect on its interception of radiation is the amount of leaf present, i.e. L. The leaf area index usually lies between 1 and 12 with typical values, of 3–4, for a good cover of planophile species such as alfalfa, but much higher values, of 5–10, for more erectophile species such as grasses and cereals or species with highly clumped leaves such as spruce trees. Much higher values of L, exceeding 20, have been reported for dense stands of rigidly erect-leaved plants such as *Gladiolus*. The distribution of leaf area within the canopy is also important, in addition to the amount of leaf present, because it defines where most of the light will be absorbed. It is, for example, very evident that there is more leaf present in some parts of the crown of a tree than in others, and in some parts of a canopy there may, of course, be large gaps with no leaves present at all.

The distribution of leaf area in space is defined by the leaf area density, the area of leaf per unit volume of canopy space. Where the leaf area density is small, there is a high probability that a beam of light will pass through a gap to the leaves lower down. Where the leaf area density is large, however, a beam of light will not be able to pass through to lower levels, most of the light being absorbed or scattered by the leaves there.

A large leaf area can be maintained as an efficient functioning system if all the leaves in the canopy are evenly illuminated at intermediate light flux densities. A small leaf area density caused, for example, by the spaces between plants, allows light to penetrate to leaves at lower levels in the canopy where it can then be absorbed. If all the light was intercepted at the top of the canopy by a large leaf area there, or by horizontal leaves, little light would get to lower levels and only a small leaf area could be sustained. On the other hand, a progressive increase in leaf area density with depth, or a change from more nearly vertical to more nearly horizontal leaves lower down leads to effective beam penetration and a more even distribution of light. In such cases, large leaf area indices can be sustained because only few leaves are then light-saturated and even the leaves at the base of the canopy receive sufficient light for physiological processes.

The three-dimensional distribution of leaves in a canopy is difficult to measure so that little quantitative information is presently available on the uniformity of distribution of the leaves, or lack of it. For this reason it is often assumed in models of light interception and tree growth, that the leaves are randomly distributed throughout the canopy volume, although in many cases this is obviously not so.

In general, leaves are aggregated or grouped, rather than distributed uniformly or randomly throughout the canopy. In a forest, for example, leaves are grouped into the crowns of the trees and, within the crowns, they may be grouped into whorls of branches. Within whorls, the leaves are most frequently grouped around the shoots. This can be seen clearly in spruce trees and other conifers but occurs to a lesser extent in other plants. These different levels of grouping have a major influence on the penetration of light through the canopy to lower levels and hence influence both light interception by the leaves and the total leaf area that can be sustained. We may arbitrarily define the maximum possible leaf area index as the value that intercepts 95% of the incident radiation (i.e. $\tau = 0.05$). If the leaves were randomly distributed, i.e. with $K = 0.5$, a leaf area index exceeding 6 could not be sustained by any species, whereas grouping allows a leaf area index of 10 or more.

In dense vegetation it is generally adequate to describe the distribution of leaf area in terms of the vertical distribution of the foliage. This has often been defined by a normal distribution or by a beta function. Where the plants are further apart, as in open stands of trees, savannah and agroforestry plantations, the overstorey foliage is confined within the crowns of the trees with substantial spaces in between. It is then necessary to define the volume, shape and position of the crowns. Within the crowns, both the vertical and radial distribution of the foliage vary and may be described by beta functions in both dimensions. While the simplest assumption is that of uniform leaf area density throughout the crown, in practice this is far from the case. The end result of combining typical vertical and horizontal distributions within the constraints of the crown shape is a three-dimensional contour map of leaf area density. For example, this may vary from 15 m^2m^{-3} in the densest part of the crown to 1 m^2m^{-3}

Fig. 2.4. A two-dimensional contour map of the distribution of leaf area density within the right-hand half of a tree crown of Sitka spruce (*Picea sitchensis*). The leaf area density at any point is obtained from $f.A/(h.r^2(z))$ where A is the total leaf area in the crown (m^2), h is crown length (m), and $r(z)$ (m) is the radius of the crown at any relative height z. (Data of Y.–P. Wang).

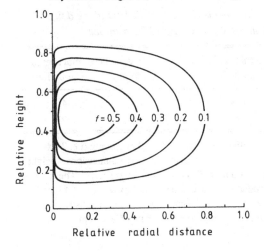

at the other extreme, with an overall average of about 2 $m^2 m^{-3}$. In some tree crowns where this has been described, the densest foliage is to be found in the upper third of the crown about one-fifth of the radial distance from the trunk (Fig. 2.4).

The consequence of such a distribution of leaf area density in arrays of tree crowns has been investigated by simulation using the radiative transfer and photosynthesis model, MAESTRO, and some results are shown in Table 2.2. A non-uniform distribution of leaf area density like that in Fig. 2.4 increases both the amount of PAR absorbed by the canopy and the amount of photosynthesis by comparison with a canopy of similar leaf area index but with a completely uniform distribution of leaf area density. This is the result of a more even distribution of absorbed PAR within the non-uniform canopy than in the uniform canopy. An even more significant consequence of a non-uniform leaf area density is that a much larger leaf area index can be supported (i.e. 9.5 rather than 6.0) than in the uniform canopy and this leads to an even larger amount of absorbed radiation and photosynthesis, with the likelihood of a corresponding increase in growth.

The display, as well as the amount and distribution of the leaves, also affects light interception, and thus another important property of the canopy is the angular presentation of the leaves. The extremes are canopies with predominantly horizontal leaves or with predominantly vertical leaves, but a wide range of leaf angles may occur between these extremes. A common case, often assumed in canopy models when information is lacking, is an angular distribution consisting of a set of facets arranged on the surface of a sphere (spherical distribution), like the rotating multi-faceted spherical mirrors which are common in discotheques. The mean leaf inclination angle in this case is 57°. Other distributions are, of course, found in practice – when they are measured – and other assumed distribution functions are possible, such as the facets on the surfaces of ellipses of different shapes, with a

Table 2.2. *The results of simulations using the model MAESTRO of the daily absorption of PAR and assimilation of CO_2 in a stand of Sitka spruce.*

L	PAR absorbed (mol tree^{-1})		CO_2 assimilated (mol tree^{-1})	
	Uniform	Non-uniform	Uniform	Non-uniform
6.0	150	184	2.7	4.1
9.5	151	224	2.4	4.9

Note: the stand has the following properties: trees on a 2.0 x 2.0 m spacing (i.e. 2500 trees ha^{-1}), all trees identical with conical crowns with the following dimensions – height of tree 10.3 m, height to base of crown 1.5 m, radius of crown at base 1.7 m. The simulations are for a sunny day in Scotland with a total quantum flux density of 45 mol m^{-2} and a beam fraction of 0.68. The non-uniform canopy has a leaf area density distribution as in Fig. 2.4. The leaf area of both canopies is 38 m^2. Data from Y.–P. Wang.

range of ratios between the vertical and radial axes (Campbell, 1986; Campbell & Norman, in this volume).

In a planophile canopy, full PAR interception is achieved by only a small leaf area index and photosynthesis by the leaves and growth by the canopy approaches PAR saturation. With a wider range of leaf inclination angles, many of the leaves will be at intermediate values of PAR, a larger leaf area index can be supported and a higher growth rate would be expected, even though absorption of PAR is the same. Simulation with MAESTRO, for the same canopy as before, indicates that a spherical leaf angle distribution, and $L = 9.5$ could result in 1.3 times the photosynthesis of a canopy with a planophile distribution and $L = 6.0$. Similar but different results of this kind will, of course, be obtained from other inclination angle distributions in other latitudes and with different proportions of beam and diffuse radiation.

Effects of environmental stresses

An environmental stress may lead to a reduction in growth through either a reduction in ε ($A \rightarrow X$, in Fig. 2.1) or through a reduction in leaf area and, hence, in interception of radiation ($A \rightarrow B$), or through both acting together or in sequence.

Shortage of water has been shown to be the cause of a reduction in ε in a number of species, for example barley (*Hordeum vulgare*) (Legg et al., 1979) and chickpea (*Cicer arietinum*) (Hughes et al., 1987). The mechanism involved includes a reduction in photosynthetic rate as a result of both stomatal closure and premature senescence of the leaves. However, for annual crops growing in areas of only moderate water shortage the major effect of water stress on growth rate is through the development and maintenance of the canopy and thus on the amount of radiation absorbed. In the barley experiments at Rothamsted, for example, withholding water for 10 weeks led to a reduction in ε of *ca* 24%, but to a reduction in radiation interception of *ca* 48% (i.e. double the effect). In contrast, the much more severe water-stress experienced by the chickpea in Syria led to a much larger reduction in ε following severe loss of leaves. In oil palm, where leaf abscission does not occur readily, ε also falls substantially in response to water stress (Squire, 1986). It seems that ε is maintained initially by reduction in leaf area where possible, but that as stress develops ε becomes progressively reduced. This is consistent with studies on photosynthesis of leaves in water stressed crops (e.g. Ludlow, 1985). Both stomatal closure and reduction in the carboxylation capability are involved. Large atmospheric water saturation deficits also cause stomatal closure and have been reported to reduce ε of pearl millet (*Pennisetum typhoides*) (Squire, Marshall & Ong, 1986).

Temperature can also affect ε as well as canopy development. In pearl millet ε varied only between 2.2–2.6 g MJ^{-1} (solar) over the average temperature range 19–31 °C (Squire et al., 1984; Mohamed, Clarke & Ong, 1988), whereas in the C_3 groundnut (*Arachis hypogaea*) ε varied from 1.0–2.1 g MJ^{-1} over the same temperature range.

Other factors that have been reported to affect ε include ozone concentration (Unsworth, Lesser & Heagle, 1984), and nutrition (Green, 1984).

The seasonal pattern of L depends strongly on the pattern of leaf initiation, emergence, expansion, senescence and death, and thus on the species under consideration. In some vegetation types, such as pasture grasses growing in a humid climate, the proportion of PAR absorbed remains relatively constant over the year and the same is true of evergreen forest with a large and relatively constant leaf area index. In these cases the annual PAR absorbed is simply the product of the annual PAR receipts and the mean absorptance. More commonly, there is a seasonal cycle of leaf area, most pronounced with annual plants but present in a wide variety of communities including tropical rain forest (Malingreau, 1986). Environmental variables which influence the time course of radiation absorptance can be broken down into those which affect the rate of leaf emergence and senescence, such as temperature, and those which affect leaf expansion, such as water and nutrient availability. Both these sets of variables can be invoked to explain the differences between genotypes that have been observed in many species, for example field beans (Green, Hebblethwaite & Ison, 1985) and peas (Heath & Hebblethwaite, 1985).

The seasonal pattern of radiation absorptance can be modelled either directly as a function of environmental variables, or indirectly by predicting L and then substituting the values into eqns (4) and (8). Milford et al. (1985) have described the effect of nitrogen and temperature on the leaf area index of sugar beet (Beta vulgaris) and this analysis could easily be extended to include the PAR absorptance of the crop. The advantage of considering leaf area index and density is that this allows the development of mechanistic rather than empirical relationships, since many effects of the environment can be assessed in terms of their influence on the leaf population. However, L alone, for the reasons already considered, may not be a sufficient description of canopy structure.

Conclusions

A simple model of the growth of plant stands can be created in terms of radiation receipts, the fraction of radiation absorbed by the canopy and the dry matter: radiation quotient. Both the absorption of radiation by the canopy and ε can be related to the seasonal development of the vegetation and to environmental variables such as temperature, nutrients and water. These variables may be introduced into eqn (3) explicitly as single factors that moderate the growth rate. Relationships that define these factors can be derived from physical and physiological considerations so that they contain parameters that relate in a meaningful way to processes that are, at least, partially understood (Monteith, 1977; Jarvis & Leverenz, 1983; Landsberg, 1986). Such a model can be useful in predicting the responses of crops, plantations, or natural plant communities to different environments or treatments and provides a means whereby such vegetation may be managed on a rational biological basis. The advantage of this approach is that the initial use of such a 'top-down' model does not depend on a very detailed understanding of all the processes involved. The model is based initially on empirical observation and has immediate practical application. A more detailed understanding of the underlying processes results as the model is developed into an interpretative tool.

References

Bedel, J.A. (1983). Solar energy recovered by a flat plate collector. In *Solar Radiation Data*, ed. W. Palz, pp. 170–6. London: D. Reidel Publishing Company.

Birnie, R.V., Millard, P., Adams, M.J. & Wright, G.G.(1987). Estimation of percentage ground cover in potatoes by optical radiance methods. *Research and Development in Agriculture*, **4**, 33–5.

Björkman, O. (1981). Responses to different quantum flux densities. In *Encyclopedia of Plant Physiology, New Series, Vol. 12A, Physiological Plant Ecology I*, eds. O.L. Lange, P.S. Nobel, C.B. Osmond & H. Ziegler, pp. 57–107. Berlin, Heidelberg: Springer–Verlag.

Bonhomme, R., Ruget, F., Derieux, M. & Vincourt, P. (1982). Relations entre production de matière sèche aérienne et énergie interceptée chez différents génotypes de maïs. *Compte Rendu de l'Academie des Sciences, Paris*, **294**, 393–8.

Campbell, G.S. (1986). Extinction coefficients for radiation in plant canopies calculated using an ellipsoidal inclination angle distribution. *Agricultural and Forest Meteorology*, **36**, 317–21.

Cannell, M.G.R., Milne, R., Sheppard, L.J. & Unsworth, M.H. (1987). Radiation interception and productivity of willow. *Journal of Applied Ecology*, **24**, 261–78.

Gallagher, J.N. & Biscoe, P.V. (1978). Radiation absorption, growth and yield of cereals. *Journal of Agricultural Science, Cambridge*, **91**, 47–60.

Gallo, K.P. & Daughtry, C.S.T. (1986). Techniques for measuring intercepted and absorbed photosynthetically active radiation in corn canopies. *Agronomy Journal*, **78**, 752–6.

Gosse, G., Varlet–Grancher, C., Bonhomme, R., Chartier, M., Allirand, J–M. & Lemaire, G. (1986). Production maximale de matière sèche et rayonnement solaire intercepté par un couvert végétal. *Agronomie*, **6**, 47–56.

Goudriaan, J. (1977). Crop micrometerology: A simulation study. Simulation monograph. Wageningen:PUDOC.

Green, C.F. (1984). Analysis of wheat growth in relation to husbandry and environment. *University of Nottingham PhD thesis*.

Green, C.F. (1986). A reappraisal of biomass accumulation by temperate cereal crops. *Speculations in Science and Technology*, **9**, 193–212.

Green, C.F., Hebblethwaite, P.D. & Ison, D.A. (1985). A quantitative analysis of varietal and moisture status effects on the growth of *Vicia faba* in relation to radiation absorption. *Annals of Applied Biology*, **106**, 143–55.

Heath, M.C. & Hebblethwaite, P.D. (1985). Solar radiation intercepted by leafless, semi-leafless and leafed peas (*Pisum sativum*) under contrasting field conditions. *Annals of Applied Biology*, **107**, 309–18.

Hipps, L.E., Asrar, G. & Kanemasu, E.J. (1983). Assessing the interception of photosynthetically active radiation in winter wheat. *Agricultural Meteorology*, **28**, 253–9.

Hughes, G., Keatinge, J.D.H., Cooper, P.J.M. & Dee, N.F. (1987). Solar radiation interception and utilization by chickpea (*Cicer arietinum* L.) crops in northern Syria. *Journal of Agricultural Science, Cambridge*, **108**, 419–24.

Hughes, G., Keatinge, J.D.H. & Scott, S.P. (1981). Pigeon peas as a dry season crop in Trinidad, West Indies. II Interception and utilization of solar radiation. *Tropical Agriculture, Trinidad*, **58**, 191–9.

Jackson, J.E. & Palmer, J.W. (1981). Light distribution in discontinuous canopies: calculation of leaf areas and canopy volumes above defined 'irradiance contours' for use in productivity modelling. *Annals of Botany*, **47**, 561–5.

Jarvis, P.G. (1981). Production efficiency of coniferous forest in the UK. In *Physiological Processes Limiting Plant Productivity*, ed C.B.Johnson, pp. 81–107. London: Butterworth.

Jarvis, P.G. & Leverenz, J.W. (1983). Productivity of temperate, deciduous and evergreen forests. In *Encyclopedia of Plant Physiology New Series, Vol. 12D, Physiological Plant Ecology IV*, eds. O.L. Lange, P.S. Nobel, C.B. Osmond and H. Ziegler, pp. 233–80. Berlin, Heidelberg: Springer–Verlag.

Kasanga, H. & Monsi, M. (1954). On the light transmission of leaves, and its meaning for the production of matter in plant communities. *Japanese Journal of Botany*, **14**, 304–24.

Kasim, K. & Dennett, M.D. (1986). Radiation absorption and growth of *Vicia faba* under shade at two densities. *Annals of Applied Biology*, **109**, 639–650.

Landsberg, J.J. (1986). *Physiological Ecology of Forest Production.* pp. 197. London: Academic Press.

Lang, A.R.G. (1986). Leaf area and average leaf angle from transmission of direct sunlight. *Australian Journal of Botany*, **34**, 349–55.

Lang, A.R.G., Yuequin, X. & Norman, J.M. (1985). Crop structure and the penetration of direct sunlight. *Agricultural and Forest Meteorology*, **35**, 83–101.

Lang, A.R.G., Yuequin, X. (1986). Estimation of leaf area index from transmission of direct sunlight in discontinuous canopies. *Agricultural and Forest Meteorology*, **37**, 220–43.

Ledig, F.T., Drew, A.P. & Clark, J.G. (1976). Maintenance and constructive respiration, photosynthesis and net assimilation rate in seedlings of pitch pine (*Pinus rigida* Mill.) *Annals of Botany*, **40**, 289–300.

Legg, B.J., Day, W., Lawlor, D.W. & Parkinson, K.J. (1979). The effects of drought on barley growth: models and measurements showing the relative importance of leaf area and photosynthetic rate. *Journal of Agricultural Science, Cambridge*, **92**, 702–16.

Linder, S. (1985). Potential and actual production in Australian forest stands. In *Research for Forest Management*, eds. J.J. Landsberg & W. Parson, pp. 11–51. Melbourne: CSIRO.

Ludlow, M.M. (1985). Photosynthesis and dry matter production in C_3 and C_4 pasture plants, with special emphasis on tropical C_3 legumes and C_4 grasses. *Australian Journal of Plant Physiology*, **12**, 557–72.

McCartney, H.A. (1978). Spectral distribution of solar radiation. II. Global and diffuse. *Quarterly Journal of the Royal Meterorological Society*, **104**, 911–26.

Malingreau, J.-P. (1986). Global vegetation dynamics: satellite observations over Asia. *International Journal of Remote Sensing*, **7**, 1121–46.

Mendham, N.J., Shipway, P.A. & Scott, R.K. (1981). The effects of delayed sowing and weather on growth, development and yield of winter oil-seed rape (*Brassica napus*). *Journal of Agricultural Science, Cambridge*, **96**, 389–416.

Milford, G.F.J., Pocock, T.O., Jaggard, K.W., Biscoe, P.V., Armstrong, M.J., Last, P.J. & Goodman, P.J. (1985). An analysis of leaf growth in sugar beet iv. The expansion of the leaf canopy in relation to temperature and nitrogen. *Annals of Applied Biology*, **107**, 167–78.

Mohamed, H.A., Clarke, J.A. & Ong, C.K. (1988). Genotypic differences in the temperature responses of tropical crops. III. Light interception and dry matter production of pearl millet (*Pennisetum typhoides* S. & H.). *Journal of Experimental Botany* (in press).

Monteith, J.L. (1977). Climate and efficiency of crop production in Britain. *Philosophical Transactions of the Royal Society of London, Series B*, **281**, 277–94.

Monteith, J.L. & Elston, J. (1983). Performance and productivity of foliage in the field. In *The Growth and Functioning of Leaves*, eds. J.E. Dale & F.L. Milthorpe, pp. 499–518. Cambridge University Press.

Norman, J.M. & Jarvis, P.G. (1974). Photosynthesis in Sitka spruce (*Picea sitchensis* (Bong.) Carr.). III. Measurement of canopy structure and interception of radiation. *Journal of Applied Ecology*, **11**, 375–98.

Palmer, J.W. (1980). The measurement of visible irradiance using filtered and unfiltered tube solarimeters. *Journal of Applied Ecology*, **17**, 149–50.

Penning de Vries, F.W.T., Brunsting, A.H.M. & van Laar, A.H. (1974). Products requirements and efficiency of biosynthesis, a quantitative approach. *Journal of Theoretical Biology*, **45**, 339–77.

Russell, G. & Ellis, R.P. (1988). The relationship between leaf canopy development and yield of barley. *Annals of Botany*, **113**, 357–74.

Russell, G., Ellis, R.P., Brown, J., Milbourn, G.M.M. & Hayter, A.M. (1982). The development and yield of autumn- and spring-sown barley in south east Scotland. *Annals of Applied Biology*, **100**, 167–78.

Scott, R.K. & Allen, E.J. (1978). Crop physiological aspects of importance to maximum yields in potatoes and sugar beet. In *Maximum Yield of Crops*, eds. J.K.R. Gasser & B. Wilkinson, pp. 25–30. London: HMSO.

Šesták, Z. (1981). Leaf ontogeny and photosynthesis. In *Physiological Processes Limiting Plant Productivity*, ed. C.B. Johnson, pp. 147–58. London: Butterworths.

Shuttleworth, J. & Wallace, J.S. (1985). Evaporation from sparse crops – an energy combination theory. *Quarterly Journal of the Royal Meteorological Society*, **111**, 839–55.

Sinha, S.K., Bhargava, S.C. & Goel, A. (1982). Energy as the basis of harvest index. *Journal of Agricultural Science, Cambridge*, **99**, 237–38.

Smart, R.E. (1973). Sunlight interception by vineyards. *American Journal of Enology and Viticulture*, **24**, 141–7.

Squire, G.R. (1986). *A Physiological Analysis for Oil Palm Trials*. Palm Oil Research Institute of Malaysia Bulletin, **12**, 12–31.

Squire, G.R., Marshall, B. & Ong, C.K. (1986). Development and growth of pearl millet (*Pennisetum typhoides* S. & H.) in response to water supply and demand. *Experimental Agriculture*, **22**, 289–300.

Squire, G.R., Marshall, B., Terry, A. & Monteith, J.L. (1984). Response to temperature in a stand of pearl millet. 6. Light interception and dry matter production. *Journal of Experimental Botany*, **35**, 599–610.

Steven, M.D., Biscoe, P.V., Jaggard, K.W. & Paruntu, J. (1986). Foliage cover and radiation interception. *Field Crops Research*, **13**, 75–87.

Szeicz, G., Monteith, J.L. & dos Santos, J. (1964). Tube solarimeters to measure radiation among plants. *Journal of Applied Ecology*, **1**, 169–74.

Unsworth, M.H., Lesser, V.M. & Heagle, A.S. (1984). Radiation interception and the growth of soybeans exposed to ozone in open-top field chambers. *Journal of Applied Ecology*, **21**, 1059–79.

Warren-Wilson, J. (1963). Estimation of foliage angle by inclined point quadrats. *Australian Journal of Botany*, **11**, 95–105.

Watson, D.J. (1947). Comparative physiological studies on the growth of field crops. *Annals of Botany*, **11**, 41–76.

Watson, D.J. (1958). The dependence of net assimilation rate on leaf area index. *Annals of Botany*, **22**, 37–54.

M. R. RAUPACH

3. Turbulent transfer in plant canopies

Introduction

This chapter is about turbulent transfer between a plant community and the atmosphere, especially the transfer of heat, water vapour, CO_2 and other scalar entities. We consider the way in which turbulent transfer influences the microclimate within the plant community, in particular the mean scalar concentrations, including temperature and humidity. The second section provides a brief, qualitative overview, to establish the connections between the turbulent transfer process in a plant canopy and exchange processes at both smaller scales, those of individual leaves, and larger scales, those of the entire planetary boundary layer of the atmosphere. Then, with frequent reference to the observed properties of turbulence in plant canopies, the third and fourth sections review two kinds of theory for describing turbulent transfer. The more common *Eulerian* theories consider the behaviour of the turbulent transfer process at a grid of points fixed in space. Less common, but of increasing importance, are the *Lagrangian* theories: these describe turbulent transfer by considering the statistical behaviour of the wandering blobs of fluid which actually carry the transferred entity.

Overview

Common experience shows that the air motion within a plant canopy is highly erratic and intermittent. The origin of this behaviour lies in the *planetary boundary layer* (PBL), the turbulent layer of the atmosphere which extends from the ground to a height of the order of a kilometre (within a factor of three or so). Turbulence in the daytime PBL is generated mechanically by the action of wind shear, which converts the kinetic energy of the mean flow into kinetic energy of turbulence, and thermally by buoyant convection, which converts potential energy into kinetic energy. Both mechanisms result in large eddies which occupy the whole depth of the PBL and keep the bulk of the PBL well mixed. The large eddies are themselves unstable: they transfer their kinetic energy to a cascade of successively smaller and smaller eddies, leading eventually (when the eddies become so small that they are smeared out by molecular viscosity) to dissipation of the kinetic energy as heat. McNaughton (this volume) discusses the well-mixed part of the daytime PBL, and its relationship with the energy balance at the surface.

Near the bottom of the PBL (say, at heights less than a tenth of the PBL depth) the large eddies are distorted by the surface into nearly horizontal motions with little vertical component and therefore little ability to promote vertical mixing. Here, the main vertical mixing is done by smaller eddies in the cascade. With approach to the ground, successively smaller eddies dominate the vertical transfer; in fact (for conditions of neutral thermal stability) the dominant vertical length scale of the eddies is directly proportional to the height and the mean velocity profile is logarithmic with height. The layer where this happens is called the inertial sublayer (Tennekes & Lumley, 1972). However, the inertial sublayer does not extend quite to the surface: beneath it lies the roughness sublayer, the region where the turbulence structure depends explicitly on the details of the (rough) surface, because of the spatially distributed pattern of momentum absorption (Raupach, Thom & Edwards, 1980). The flow in the immediate environment of a plant canopy is that of the roughness sublayer, which typically extends to about two canopy heights above the underlying ground. Clearly, turbulence within and just above the canopy incorporates a very wide range of eddy sizes; in other words, the eddy spectrum (the distribution of eddy kinetic energy with eddy size) is broad and continuous.

To form a qualitative picture of turbulence in the canopy environment, it is useful to divide the continuous eddy spectrum into three ranges, corresponding to eddies with vertical length scales (L_w) much larger than the canopy height (h_c), comparable with h_c and much smaller than h_c. The large eddies, with $L_w \gg h_c$, appear in the canopy environment to be nearly two-dimensional motions which maintain little vertical mixing, simply causing the air in the canopy to slosh about. (They organise and modulate smaller eddies, however). The eddies with $L_w \sim h_c$ make up the vertically energetic turbulence which dominates vertical transfer. The small eddies, with $L_w \ll h_c$, are produced not only by the eddy cascade but also, more importantly, by the turbulent wakes which develop downstream of leaves and stems; these eddies are short-lived (because of their small size) and too small to participate much in vertical transfer.

The most important eddies for vertical transfer are those with $L_w \sim h_c$. They are strikingly visualised by the wind waves or 'honami' which cross fields of grass or cereal on windy days; in fact, it was by studying honami that Finnigan (1979a,b) obtained a description of these motions. In near-neutral conditions, they consist mainly of energetic, intermittent, downward-moving gusts, with vertical length scales of the order of the canopy height, and horizontal length scales of several canopy heights; between the gusts there are more quiescent, slower-moving updraughts. As each gust passes through the canopy, the drag exerted by leaves and stems rapidly depletes the momentum and kinetic energy of the gust, converting the kinetic energy to small-scale wake turbulence which decays quickly by viscous smearing. The gusts dominate momentum transfer from the air to the canopy, because the canopy drag force (or momentum sink) is approximately proportional to the square of the local instantaneous wind speed, and is therefore weighted strongly towards gusts. The phenomenon of honami in flexible canopies takes place as gusts sweep over the canopy, bending patches of stalks before them; after a gust passes, the stalks oscillate for a few cycles at their natural frequency, creating the impression of wave motion.

This general picture of momentum transfer applies whether honami is observed or not. It is supported by a substantial and growing body of evidence, based on conditional analysis of turbulence signals, short-time profile analysis, space–time correlations and flow visualisation (for recent reviews see Finnigan & Raupach, 1987, and Raupach, Coppin & Legg, 1986). However, it is important to note that recent evidence (Coppin, 1985) suggests that scalar transfer is not as closely linked to the gusts as momentum transfer, especially in unstable conditions – although the scalar transfer is still dominated by the same class of eddies, those with $L_w \sim h_c$.

To complete this overview, we briefly consider transfer processes on the scale of individual leaves. Momentum transfer at the leaf surface takes place by a combination of pressure and viscous (molecular) forces, the pressure force usually being dominant (Thom, 1971). On the other hand, scalar transfer at the leaf surface takes place by molecular diffusion alone. Consider, for example, a leaf emitting water vapour: molecular diffusion carries the water vapour from the substomatal cavities, through the stomatal pores to the leaf surface and then through a thin *leaf boundary layer* (LBL), with a thickness of the order of a millimetre or less, to the turbulent air flow within the canopy. Only then does turbulent transfer take effect to spread the water vapour away from the leaf, into the flow above the canopy, and ultimately into the well-mixed bulk of the PBL. The state of the LBL is almost always laminar, so that the dominant transfer mechanism there is molecular rather than turbulent. However, the leaf is exposed to a highly turbulent canopy flow, and so the LBL is extremely unsteady (Finnigan & Raupach, 1987).

It is common to parameterise the molecular diffusion process through the LBL by a *boundary-layer resistance* r_b (or conductance $g_b = r_b^{-1}$) and the diffusion process through stomata by a *stomatal resistance* r_s (or conductance $g_s = r_s^{-1}$). For a scalar entity with concentration c per unit mass of air, and flux density F_0 at the leaf surface, r_b and r_s are defined by

$$F_0 = \frac{\rho(c_0 - c)}{r_b} = \frac{\rho(c_i - c_0)}{r_s} = \frac{\rho(c_i - c)}{r_s + r_b} \, , \qquad (1)$$

where ρ is the air density and c_i, c_0 and c denote concentrations in the substomatal intercellular space, at the leaf surface and in the turbulent air just outside the LBL, respectively. For the present discussion, we regard r_b, r_s and c_i as given, so that eqn (1) forms a relationship between the flux F_0 and the concentration c outside the LBL. This enables a precise statement of the problem of scalar turbulent transfer in the canopy environment to be made: how does the turbulent transfer process disperse the scalar from individual leaves in the canopy, thereby determining the distribution through the canopy of c, and thence, via eqn (1), the flux density F_0?

What is being assumed by taking r_b, r_s and c_i as given? To a rough approximation, r_b is determined by leaf size and wind speed outside the LBL, though the effect of turbulence is not yet certain (Finnigan & Raupach, 1987). The physiological state of the leaf determines r_s, while c_i is determined by biochemical feedback mechanisms in the case of CO_2 transfer (Wong, Cowan & Farquhar, 1979) and by the leaf energy balance in the case of coupled heat and water vapour transfer. In the latter case, eqn (1) is usually replaced by the leaf energy balance equation and the leaf combination equation (Monteith, 1973), but this complication does not affect the essence of the present discussion.

With this background, we turn now to Eulerian and Lagrangian theories for scalar turbulent transfer. We do not consider explicitly the relationship between F_0 and c that is implied by eqn (1), but instead regard F_0 as given. Since F_0 for an individual leaf can be thought of as specifying a scalar source strength, the problem is: what is the concentration field produced by dispersion from a given distribution of sources through the canopy? The term 'source' will be understood to include both positive and negative F_0, that is, both sources and sinks.

Eulerian theory for turbulent transfer of scalars

The starting point for all theories of scalar transfer is the conservation law for the mass of the scalar entity. This is expressed (Landau & Lifshitz, 1959, section 53) by the equation:

$$\frac{\partial c}{\partial t} + u_i \frac{\partial c}{\partial x_i} = k_c \frac{\partial^2 c}{\partial x_i \partial x_i} ,$$
(2)

where $u_i(\mathbf{x},t) = (u_1,u_2,u_3)$ is the instantaneous velocity vector, $x_i = \mathbf{x} = (x_1,x_2,x_3)$ the position vector, t is time, k_c the molecular diffusivity for the scalar, and $c(\mathbf{x},t)$ the instantaneous concentration of scalar. We use the tensor summation convention, by which an index repeated within a term is summed over its possible values, 1,2, and 3. The corresponding conservation equation for the mass of air is the continuity equation for an incompressible fluid,

$$\frac{\partial u_i}{\partial x_i} = 0 .$$
(3)

Because the flow is turbulent, both c and u_i fluctuate rapidly and randomly. To produce equations in quantities which vary smoothly and deterministically, an averaging operation is necessary. In practice, time averages are usually used, with the assumption that the flow is statistically steady. The averaged equations are obtained by splitting each instantaneous quantity into an average (denoted by an overbar) and a fluctuation about the average (denoted by a prime), so that

$$u_i = \bar{u}_i + u_i' , \quad c = \bar{c} + c' .$$
(4)

This is called the Reynolds decomposition. Substituting eqns (4) and (3) into eqn (2) and averaging the entire equation, we obtain

$$\frac{\partial c}{\partial t} = -\frac{\partial}{\partial x_i}\left(\bar{u}_i\,\bar{c} + \overline{u_i'c'}\right) - k_c\frac{\partial \bar{c}}{\partial x_i}) \ . \tag{5}$$

The term on the left is the rate of increase of $\bar{c}$ in an infinitesimal volume. This balances a negative flux divergence, or net flow of scalar into the volume. There are three fluxes: an advective flux $\bar{u}_i\,\bar{c}$, an eddy flux $\overline{u_i'c'}$, given by the covariance of u_i and c, and a molecular flux $-k_c\partial\bar{c}/\partial x_i$.

In a plant canopy, it is necessary to average not only at a point by time or ensemble averaging, but also in space, over thin horizontal slices through the canopy, to remove the complex and irregular spatial variability introduced by the plant elements. Spatial or volume averaging is discussed in detail by Finnigan (1985), who catalogues the extra terms that it introduces into the equations. In the case of eqn (5), the main additional term is a scalar source density: this term equals the surface integral, over all solid surfaces in the averaging volume, of the molecular flux of scalar at those surfaces. The other additional term arising from spatial averaging is the 'dispersive flux' term, which is the counterpart for spatial variations of the eddy flux $\overline{u_i'c'}$ (which arises from temporal fluctuations at a single point). Current experimental evidence suggests that this term is small (Raupach, Coppin & Legg, 1986), and it is neglected here. We also neglect molecular fluxes in comparison with eddy fluxes, everywhere except at solid surfaces. The scalar equation after both time and spatial averaging then becomes

$$\frac{\partial \bar{c}}{\partial t} = -\frac{\partial}{\partial x_i}(\bar{u}_i\,\bar{c} + \overline{u_i'c'}) + S(\mathbf{x},t) \ , \tag{6}$$

where $S(\mathbf{x},t)$ is the scalar source density, and where the overbar denotes both spatial *and* time averaging. Molecular fluxes appear in eqn (6) entirely through $S(\mathbf{x},t)$.

Let us suppose that the wind field ($\bar{u}_i$) and source distribution (S) are known, and try to solve eqn (6) for $\bar{c}$. The difficulty is that an unknown second moment, $\overline{u_i'c'}$, has appeared in the equation for $\bar{c}$, a first moment. Hence, the equation remains unclosed and cannot yield an explicit solution for $\bar{c}$. This is an example of the infamous closure problem of turbulence (Tennekes & Lumley, 1972), whereby an equation for any moment of a fluctuating quantity contains higher-order moments, which are extra unknowns. Equations for these higher-order moments contain unknown moments of higher order still, and so on *ad infinitum*. To close and solve the set of equations, one must write additional equations based on extra hypotheses.

The simplest way of closing eqn (6) is to postulate a relationship expressing $\overline{u_i'c'}$ in terms of $\bar{u}_i$ and $\bar{c}$. This approach is called 'first-order closure', because the only equation to be solved is one for the first moment $\bar{c}$. The usual model for $\overline{u_i'c'}$ is a gradient-diffusion hypothesis, or K-theory:

$$\overline{u_i'c'} = -K\frac{\partial \bar{c}}{\partial x_i} \ , \tag{7}$$

where K is an eddy diffusivity analogous to the molecular diffusivity k_c, but several orders of magnitude larger. However, there are serious problems with eqn (7) in the canopy environment, both on theoretical and on practical grounds. A well-known statement of the theoretical objection was given by Corrsin (1974): eqn (7) is justifiable only if the length scale of the turbulent motions which maintain the flux $\overline{u_i'c'}$ is much smaller than the length scale of changes in the mean gradient $\partial\overline{c}/\partial x_i$. Let us see how this criterion performs in real canopies.

Fig. 3.1 shows some wind data in two canopies, a wind-tunnel model (Raupach, Coppin & Legg, 1986) and a corn (*Zea mays*) canopy (Wilson *et al.*, 1982). The data are given in meteorological notation: the wind vector is written as (u,v,w) and the position vector as (x,y,z), with x along the mean wind, z upward and $z = 0$ the ground. Heights and turbulent length scales are normalized by the canopy height h_c (60 mm for the model, 2.2 m for the corn) and velocities by the friction velocity u_* (defined by $\tau = \rho u_*^2$, where τ is the downward momentum flux, $-\rho\overline{u'w'}$, in the constant-flux layer above the canopy). The model canopy was quite sparse (leaf area index $\lambda = 0.23$) whereas the corn was much denser ($\lambda = 3$); this accounts for the fact that $\overline{u}$ and the turbulence statistics σ_u, σ_w and $\overline{u'w'}$ are all larger within the model than within the corn canopy. The $\overline{u'w'}$ profiles show that in both canopies, most of the momentum is absorbed in the upper half of the canopy

In both of these experiments, the turbulence length scale L_w was obtained from

$$L_w(z) = \overline{u}(z) \int_0^\infty r_{Eww}(t)\, dt \ , \tag{8}$$

where

$$r_{Eww}(t) = \overline{w'(s)\, w'(s + t)} / \overline{w'^2}$$

is the Eulerian (fixed-point) autocorrelation function of w. This length scale is an approximate measure, obtainable from a record of velocity fluctuations at a single point, of the distance in the x direction over which fluctuations of w are correlated at a

Fig. 3.1. Profiles of mean streamwise wind speed $\overline{u}$, streamwise and vertical velocity standard deviations σ_u and σ_w, kinematic momentum flux $-\overline{u'w'}$, and vertical turbulent length scale L_w, in two canopies: a wind-tunnel model of height $h_c = 60$ mm (Raupach, Coppin & Legg, 1986) and a corn canopy with $h_c = 2.2$ m (Wilson *et al.*, 1982). Velocities are normalised with the friction velocity u_*.

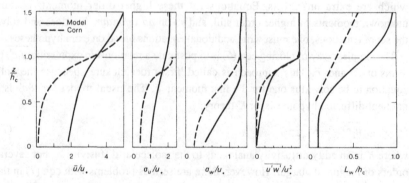

single instant of time; in other words, it is a direct measure of the eddy length scale associated with vertical velocity fluctuations. Fig. 3.1 shows that in the upper part of the canopy, where most of the momentum is absorbed, L_w is a substantial fraction of h_c; at $z/h_c = 0.8$, for example, L_w is $0.42h_c$ for the model and $0.22h_c$ for the corn. (The difference probably stems from the difference in leaf area index). More importantly, L_w is comparable with the height interval over which $\partial \bar{u}/\partial z$ changes substantially. This interval is given by $|\partial \bar{u}/\partial z| |\partial^2 \bar{u}/\partial z^2|^{-1}$ which is $0.46h_c$ for the model and $0.25h_c$ for the corn, again at $z/h_c = 0.8$. We conclude that Corrsin's length scale criterion for a K-theory is not satisfied for momentum transfer in canopies. This finding agrees with our earlier qualitative overview of the mechanism of momentum transfer to a canopy, where it was argued that momentum transfer is dominated by eddies with vertical length scales comparable with h_c.

The situation for scalar transfer is similar, as is demonstrated in practice by the existence of counter-gradient fluxes (i.e. negative K) in pine forests. Fig. 3.2, from Denmead & Bradley (1985), shows simultaneous measurements of profiles and vertical fluxes at two levels, for heat, water vapour and CO_2 transfer. Above the canopy, gradients and fluxes have opposite signs in each case, so K is positive. However, there is a well-defined region below the canopy crown where both the heat and CO_2 fluxes take the same sign as the corresponding gradients, implying negative K values. There is a sizeable water vapour flux in the trunk space, resulting from soil evaporation, but a negligible gradient, implying an infinite K. Observations like these invalidate efforts to infer source distributions within the canopy by differentiating measured profiles (Brown & Covey, 1966), as there is no rational way to choose the necessary K value. We show later how a Lagrangian theory can explain counter-gradient fluxes.

Fig. 3.2. Profiles of average temperature $\bar{\theta}$, mass fraction of water vapour $\bar{q}$ and volume fraction of CO_2 $\bar{c}$ in Uriarra forest over one hour, with simultaneously measured fluxes of sensible heat H, latent heat λE (both in W m^{-2}) and $CO_2 F_c$ (in mg m^{-2} s^{-1}) at two levels. Reproduced from Denmead & Bradley (1985) with the permission of D. Reidel Publishing Company.

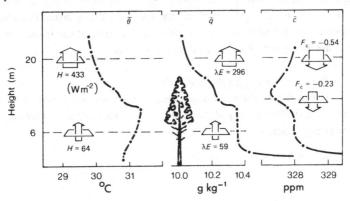

It is apparent that the gradient-diffusion hypothesis, eqn (7), has serious deficiencies in the canopy environment. In an effort to overcome these deficiencies, some workers have investigated 'higher-order closure models'. Unlike the first-order closure model formed by eqns (6) and (7), these models include conservation equations for the second moment $\overline{u_i'c'}$ and possibly for higher moments as well. The closure problem means that the set of equations contains unknown terms, so additional closure hypotheses are always needed. In higher-order closure models, these hypotheses are complicated and difficult to test. However, Deardorff (1978) showed that in order to obtain correct results from second- and third-order closure models for scalar transfer, a gradient-diffusion hypothesis similar to eqn (7) must be introduced at the highest order of the model. This suggests that resort to higher-order closure does not obviate the difficulties caused by eqn (7).

There is not space here for detailed discussion of higher-order closure models, or of work with them in the canopy environment. The reader is referred to Donaldson (1973), Launder (1976) and Lumley (1978) for general reviews; to Wilson & Shaw (1977) and Finnigan (1985) for work in canopies; and to Raupach (1987) for a critique. Let us summarize with the statement that higher-order closure does not of itself eliminate the objections to the gradient-diffusion hypothesis, since this hypothesis always enters at some point or other.

Lagrangian theory for turbulent transfer of scalars

All transfer or dispersion theories are based ultimately upon conservation laws. The terms *Eulerian* and *Lagrangian* in the headings to this and the previous section refer to the coordinate framework in which the equations expressing the laws are considered: Eulerian theories solve the equations in a grid of infinitesimal control volumes which are fixed in space so that the fluid flows through them, while Lagrangian theories use infinitesimal control volumes which move with the fluid, so that there is no flow through their walls. The Lagrangian viewpoint readily offers an insight into the process of turbulent dispersion which is unavailable in an Eulerian context, namely, the dominant role played by persistence of turbulent fluid motion. This turns out to have great significance for turbulent transfer in canopies, as explained in the rest of this chapter. Before turning to plant canopies, however, we describe the basis for the Lagrangian theory in general.

Basis for Lagrangian dispersion theory

Let us follow the motion of a particular infinitesimal particle of fluid – a material particle. The velocity of the particle, $U_i(t)$, is a *Lagrangian velocity*, and the position of the particle is

$$\mathbf{X}(t) = X_i(t) = x_{0i} + \int_{t_0}^{t} U_i(s) \, ds \; , \tag{9}$$

where x_{0i} is the particle's starting position at $t = t_0$. (The vector notation $\mathbf{X}$ is more convenient than the tensor notation X_i when the position is an independent variable in a function, such as $c(\mathbf{X},t)$.) The Lagrangian velocity is the same as the Eulerian velocity at the position of the particle:

$$U_i = \frac{dX_i}{dt} = u_i(\mathbf{X}(t),t) \ . \tag{10}$$

The rate of change of scalar concentration in the fluid particle is

$$\frac{dc(\mathbf{X}(t),t)}{dt} = \frac{\partial c}{\partial t} + u_i \frac{\partial c}{\partial x_i} \ , \tag{11}$$

where the equality follows from the chain rule for differentiation and eqn (10). This rate of change is called a material or total derivative, and is denoted dc/dt in contrast to the partial derivative $\partial c/\partial t$ which denotes the rate of change of c at a fixed point $\mathbf{x}$. With the aid of eqn (11), the scalar conservation equation, eqn (2), can be written in particle-following or Lagrangian form:

$$\frac{dc}{dt} = k_c \frac{\partial^2 c}{\partial x_i \partial x_i} \ , \tag{12}$$

where the right-hand side is evaluated at $\mathbf{x} = \mathbf{X}(t)$. Hence, the fluid particle changes its scalar concentration only by molecular diffusion.

The Lagrangian theory for turbulent dispersion makes direct use of eqn (12), in the following way. Firstly, scalar sources – which in reality are molecular scalar fluxes at solid boundaries – are represented by a source density function $S(\mathbf{x},t)$, with the dimensions of mass of scalar per unit time per unit mass of air. At source locations S is non-zero, but $S = 0$ everywhere else. Secondly, molecular diffusion is assumed to be negligible, so that $dc/dt = 0$, except at source locations where its role is described by $S(\mathbf{x},t)$. These idealisations are analogous to those in the spatially averaged Eulerian conservation equation, eqn (6). The conservation equation for a fluid particle, or Lagrangian conservation equation, now becomes:

$$\frac{dc}{dt} = S(\mathbf{X}(t),t) \ . \tag{13}$$

To find the concentration field resulting from a particular source distribution $S(\mathbf{x},t)$, eqn (13) must be integrated along particle paths. The first step in doing this is to consider the complete source to be a superposition of numerous *elementary sources*, each of which approximates an instantaneous point release of unit mass of scalar at some point $\mathbf{x}_0$ and time t_0. A single elementary source produces an initial concentration, c_0, in a tiny blob of fluid centred on $\mathbf{x}_0$ at time t_0 (note that $c_0 \to \infty$ as the blob size goes to zero). Thereafter, the resulting concentration field c_e obeys

$$\frac{dc_e}{dt} = 0 \ , \ t > t_0 \ , \tag{14}$$

so the blob of fluid, as it wanders, preserves its initial concentration and acts as a *marked fluid particle*. Since the full conservation equation, eqn (13), is linear, the complete concentration field c(x,t) is given by a superposition with weighting S of the elementary concentrations:

$$c(\mathbf{x},t) = \iint c_e(\mathbf{x},t|\mathbf{x}_0,t_0)\, S(\mathbf{x}_0,t_0)\, d\mathbf{x}_0\, dt_0 \,. \tag{15}$$

Here $c_e(\mathbf{x},t|\mathbf{x}_0,t_0)$ denotes the *elementary concentration field*, at x and t, resulting from an elementary source at $\mathbf{x}_0$ and t_0. The space integral covers the whole fluid and the time integral runs from $-\infty$ (when $c = 0$ everywhere) to the time t.

The next step is to note that, because of the random nature of the turbulence, we can only ever predict *statistical* properties of the flow. These are always defined, in the Lagrangian theory, by the operation of *ensemble averaging* (see, for example, Monin & Yaglom, 1971). For a randomly varying quantity, such as $c(\mathbf{x},t)$, the ensemble average $[c](\mathbf{x},t)$ (denoted by square brackets) is the arithmetic average of the values taken by $c(\mathbf{x},t)$ in a large number of independent experimental realisations of the flow, each subject to the same initial and boundary conditions, and differing only in the random variations from one realisation to another. If $c(\mathbf{x},t)$ is statistically steady in time, as happens for dispersion from a steady, continuous source, then the ensemble average is the same as the more familiar time average.

The ensemble average of the elementary concentration field $c_e(\mathbf{x},t|\mathbf{x}_0,t_0)$ is particularly important: for any realisation, c_e is zero everywhere except at the position $\mathbf{X}(t)$ of the marked fluid particle, where it equals the initial concentration c_0. Furthermore, the integral of c_e over x is always 1. These properties mean that the ensemble average of c_e is equal to the conditional probability density function $P(\mathbf{x},t|\mathbf{x}_0,t_0)$ for the position $\mathbf{X}(t)$ of a particle starting from $\mathbf{x}_0$ at time t_0, i.e:

$$[c_e](\mathbf{x},t|\mathbf{x}_0,t_0) = P(\mathbf{x},t|\mathbf{x}_0,t_0) \,. \tag{16}$$

This conditional probability density function is called the *transition probability* of $\mathbf{X}(t)$. It is defined thus: $P d\mathbf{x}$ is the probability that a fluid particle's position $\mathbf{X}(t)$ lies within the tiny volume element dx centred on x, given that $\mathbf{X}(t_0) = \mathbf{x}_0$. Fig. 3.3 illustrates the concept. Taking the ensemble average of eqn (15) and using eqn (16), we get

$$[c](\mathbf{x},t) = \iint P(\mathbf{x},t|\mathbf{x}_0,t_0)\, S(\mathbf{x}_0,t_0)\, d\mathbf{x}_0\, dt_0 \,, \tag{17}$$

which is the solution of the Lagrangian conservation equation, eqn (13). To evaluate $[c]$ using eqn (17), we need a model for the transition probability P, which is a statistic of the fluid motion only. The model for P plays a role in the Lagrangian theory which is equivalent to that of closure assumptions in the Eulerian theory.

Before discussing P, one other property of the ensemble average should be mentioned. The assumption leading to eqn (13), that molecular diffusion is negligible except at sources, appears crude. It is certainly wrong for individual realisations, as it leads to an unrealistic probability density function for c_e which allows only two values, $c_e = c_0$ in a marked particle and $c_e = 0$ everywhere else. However, it can be shown that the ensemble averages $[c_e]$ and $[c]$ are not seriously affected by neglecting molecular diffusion, and in fact are exact in the limit of infinite Péclet number Pe = UL/k_c, where L is a turbulence length scale and U a turbulence velocity scale (Monin & Yaglom, 1971).

Modelling the Transition Probability P

We need to know the transition probability P in order to find the mean concentration produced by a given source distribution, using eqn (17). It is worth restating the significance of the transition probability: P is the probability density that a fluid particle will end up at position x at time t, given that it started from position x_0 at time t_0. This is the same as the ensemble-averaged concentration field from an elementary source (instantaneous point source of unit mass of scalar) at (x_0, t_0). Moreover, P is a property of the fluid motion, or velocity field, only. The aim of this section is to show how P can be found from velocity statistics.

For *steady, homogeneous* turbulence, P can be expressed analytically. We consider dispersion in one dimension only, writing $Z(t)$ for the position of a fluid particle and $W(t) = dZ/dt$ for its Lagrangian velocity. The Eulerian velocity, $w(z,t)$, is assumed to have zero mean, a known variance $[w'^2]$ and a Gaussian distribution. Lagrangian and Eulerian velocity moments are always equal in steady, homogeneous turbulence (Tennekes & Lumley, 1972), so that

$$\left. \begin{array}{c} [w] = [W] = 0 \\[2mm] [w'^2] = [W'^2] = \sigma_W{}^2 \end{array} \right\} \tag{18}$$

where σ_W denotes the Lagrangian velocity standard deviation.

Fig. 3.3. Definition of the transition probability P. Only one component of the vector position is shown.

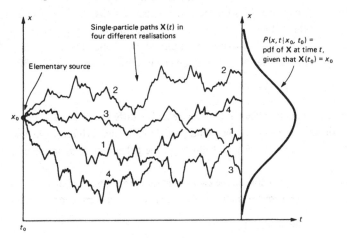

If an elementary source releases unit mass of scalar at $z = 0$ and $t = 0$, then the mean position of the cloud $[Z](t)$, is always zero. The width of the cloud, $\sigma_Z(t)$, is defined by

$$\sigma_Z^2(t) = [(Z(t) - [Z](t))^2] \ ,$$

and, by the well-known kinematic theorem of Taylor (1921), obeys

$$\frac{d\sigma_Z^2}{dt} = 2\sigma_W^2 \int_0^t r_L(s) \ ds \ , \tag{19}$$

where $r_L(s)$ is the Lagrangian velocity autocorrelation function:

$$r_L(s) = [W'(t) \ W'(t + s)] / [W'^2] \ . \tag{20}$$

The most important property of $r_L(s)$ is its integral time scale

$$T_L = \int_0^\infty r_L(s) \ ds \ , \tag{21}$$

which measures the characteristic persistence time of a fluid particle's velocity, $W(t)$. From eqn (19), it is seen that the dispersion takes place in two different fashions depending on whether $t \ll T_L$ or $t \gg T_L$. These two asymptotic limits are called the *near field* and the *far field*, respectively. In the near field, r_L is nearly 1 and the integral in eqn (19) takes the value t. One more integration of eqn (19) gives:

$$\sigma_Z = \sigma_W t \quad \text{(near field, } t \ll T_L) \ . \tag{22}$$

In the far field, the integral in eqn (19) approaches T_L, so that

$$\sigma_Z = \sigma_W (2T_L t)^{1/2} \quad \text{(far field, } t \gg T_L) \ . \tag{23}$$

Thus, the near field is the region near the elementary source where $\sigma_Z \sim t$ and a marked fluid particle moves with little change in its initial velocity, $W(0)$. In the far field the particle can be thought of as undergoing a random walk with time steps much larger than T_L, because W is uncorrelated between steps; the result is a classical diffusion process with $\sigma_Z \sim t^{1/2}$. Classical diffusion from a point source into a homogeneous medium is described by a diffusivity K such that $\sigma_Z = (2Kt)^{1/2}$ (e.g. Csanady, 1973); hence, far-field dispersion can be described by the eddy diffusivity

$$K_f = \sigma_W^2 T_L \ . \tag{24}$$

However, K_f is not applicable in the near field. The differences between near-field and far-field dispersion occur because fluid particle velocities exhibit persistence over a substantial time scale, T_L. Fig. 3.4 illustrates the behaviour of σ_Z in the near and far fields.

To interpolate $\sigma_Z(t)$ between the two asymptotic limits, an explicit form for $r_L(s)$ is needed. An excellent approximation is the exponential autocorrelation function

$$r_L(s) = \exp(-|s|/T_L) \ , \tag{25}$$

which combines with eqn (19) to give:

$$\sigma_Z^2(t) = 2\sigma_W^2 T_L^2(t/T_L - 1 + \exp(-t/T_L)) \ . \tag{26}$$

This fully specifies the standard deviation of P. The shape of P is Gaussian: this is assured for $t \ll T_L$ by the Gaussian distribution of the Eulerian velocity, and for $t \gg T_L$ by the central limit theorem. Therefore, the complete specification of P in steady, homogeneous turbulence is:

$$P(z,t|0,0) = \frac{1}{\sigma_Z\sqrt{2\pi}}\exp(-z^2/2\sigma_Z^2) \ , \tag{27}$$

with $\sigma_Z(t)$ given by eqn (26). This formula was first given by Batchelor (1949).

In inhomogeneous turbulence, no such simple analytic formula for P is available. Instead, resort must be made to the postulate that $W(t)$ is a *Markov process*. The random process (or random function of time) $W(t)$ is Markovian if, once W is specified at some time t_0, its subsequent behaviour at $t > t_0$ is completely determined by $W(t_0)$, and is also statistically independent of its history prior to t_0. This concise but somewhat cryptic general definition ensures that $W(t)$ is a process with persistence, but of a rather special and limited kind. In fact, if $W(t)$ is also constrained to be stationary and to have a Gaussian probability density function, then it has the autocorrelation function of eqn (25), the transition probability of eqn (27) and is precisely the Lagrangian velocity that we have described for stationary homogeneous turbulence. The statistical correspondence between Lagrangian velocity and a Markov process is exact, however, only in the limit of infinite Reynolds number $Re = \sigma_W^2 T_L/\nu$ where ν is the kinematic viscosity of air (Sawford, 1984). The inexactness at finite Re is that the acceleration dW/dt has an infinite variance for a

Fig. 3.4. The spread in time of a scalar released from an instantaneous point source into homogeneous turbulence, showing the near field and the far field.

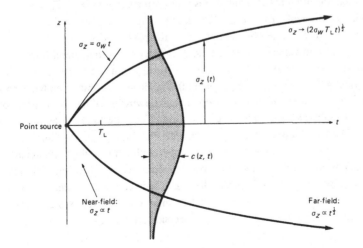

Markov process, which is clearly not possible in a real flow. However, the acceleration time scale T_K is much less than T_L, the two scales being related by

$$T_K/T_L \sim \text{Re}^{-1/2} \tag{28}$$

(Tennekes & Lumley, 1972, p. 279). Except at very short times comparable with T_K, the inexactness in the Markov model for W is negligible.

To find P in inhomogeneous turbulence, one writes a linear stochastic differential equation for the Markov process $W(t)$. Stochastic differential equations are treated in general by Arnold (1974), van Kampen (1981) and, in the context of turbulent dispersion, by Lamb (1980) and Durbin (1983). Determining the exact form of the equation for $W(t)$ is not easy; the difficulty is that σ_W and T_L are both functions of z and t, and change for a particle as it wanders through the flow. The equation must be solved numerically by random number methods, and P found by laboriously building up a large number of particle trajectories. By now, the literature on this 'random-flight method' is becoming extensive, though some of the early work was guided more by intuition than sound analysis. Two recent papers which contribute to correcting this situation are by Thomson (1984) and van Dop, Nieuwstadt & Hunt (1985), where earlier references can be found. Work specifically related to plant canopies is described by Wilson, Thurtell & Kidd (1981a,b), Legg & Raupach (1982) and Wilson, Legg & Thomson (1983).

Application in plant canopies

The turbulent transfer of heat, water vapour or CO_2 through a canopy is, in fact, superimposed scalar dispersion from a large number of approximately point or elementary sources, the individual leaves. With increasing travel time t, the plume of dispersing scalar from each leaf passes through both a near field ($t \ll T_L$, $\sigma_Z \sim t$) and a far field ($t \gg T_L$, $\sigma_Z \sim t^{1/2}$). Relatively, how important are the two regimes? The gradual transition between the two is centred at a travel time $t = T_L$, or a distance downwind of the leaf $\Delta x = \bar{u}T_L$, where $\bar{u}$ is the mean wind speed. In canopies, T_L is about $0.3h_c/u_*$ (a value obtained by Legg, Raupach & Coppin (1986) using eqn (28) and measured values of the far-field diffusivity K_f from a lateral line source of heat, in the model canopy of Fig. 3.1). A typical mean wind speed in the upper canopy is $\bar{u} \approx 3u_*$, so $\Delta x = \bar{u}T_L$ is about $0.9h_c$. Hence, the transition between near and far fields is centred about one canopy height downwind of the elementary source – a considerable distance. Therefore, near-field effects are likely to be important in canopies. This Lagrangian conclusion parallels the finding that the Eulerian length scale L_w is significant compared with the canopy height: in fact, the Lagrangian length scale $\sigma_W T_L$ is about $0.4h_c$ in the upper part of the model canopy of Fig. 3.1, and very close to L_w. Hence, the significance of the near field in canopies is directly linked with the dominance of vertical transfer by eddies with vertical length scales comparable with the canopy height, as discussed in the second section.

To see the consequences of the superposition of near-field and far-field dispersion, it is instructive to begin by neglecting the inhomogeneity of the turbulence within and above the canopy, as we can then apply the simple and powerful analytic results of eqns (26) and (27). Of course, canopy turbulence is not homogeneous, as Fig. 3.1 shows, so the results of a homogeneous-turbulence calculation cannot be construed as an exact, quantitative model. However, the physical processes emerge clearly. The following discussion is condensed from Raupach (1987).

Consider a uniform, homogeneous wind, of speed $\bar{u}$ in the x-direction and with given Lagrangian statistics σ_W and T_L. The plane $z = 0$ represents the ground. Starting at $x = 0$, the wind blows through a "canopy" of height h_c, and consisting of a scalar source with density $S(z)$, uniform in x and y downstream of $x = 0$. There is no scalar flux through the ground, which is ensured in this model by placing an image scalar source below $z = 0$. Fig. 3.5 shows the situation. We emphasise that the 'canopy' is nothing but a spatially extensive source distribution, and the 'ground' a plane of symmetry; neither has any effect on the homogeneous wind field.

Eqns (17), (26) and (27) can be used to calculate the concentration field, $\bar{c}$, at any distance x into the canopy. (Since the concentration field is steady, the ensemble average $[c]$ given by eqn (17) is equal to the more conventional time average $\bar{c}$.) By similar methods, one can find the vertical flux $\overline{w'c'}$. Fig. 3.6 shows calculated profiles of $\bar{c}$ and $\overline{w'c'}$, and also the implied vertical diffusivity $K = -\overline{w'c'}/(\partial \bar{c}/\partial z)$, with $S(z)$ a Gaussian distribution centred at $z/h_c = 0.8$ and of width 0.04. The integrated source strength or flux of c per unit ground area,

$$F_* = \int_0^{h_c} S(z) \, dz \ , \tag{29}$$

was used to normalise S and c. The time scale T_L was chosen to be $0.4h_c/\sigma_W$, and u was set at $2.5\sigma_W$ so that $\bar{u}T_L = h_c$. These relationships between σ_W, T_L, $\bar{u}$ and h_c are based on observations in the wind-tunnel model canopy of Fig. 3.1 (Raupach,

Fig. 3.5. Axes and source distribution for Lagrangian homogeneous-turbulence model for scalar dispersion in a canopy.

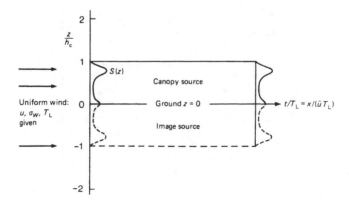

Fig. 3.6. Predicted mean concentration $\bar{c}$, vertical flux density $\overline{w'c'}$ and diffusivity $K = -\overline{w'c'}/(\partial \bar{c}/\partial z)$, in a homogeneous-turbulence canopy of height h_c, with source distribution S as shown. Normalising parameters are total scalar flux density F_*, Lagrangian vertical velocity standard deviation σ_W and far-field diffusivity $K_f = \sigma_W^2 T_L$ where T_L is Lagrangian time scale, set at $0.4h_c/\sigma_W$. Mean wind speed $\bar{u}$ is $2.5\sigma_W$ and x is distance downwind of canopy leading edge. Reproduced from Raupach (1987) with the permission of the Royal Meteorological Society.

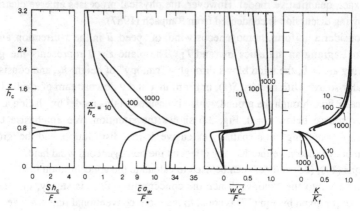

Fig. 3.7. Measured profiles of $\bar{c}/c_*$ and $\overline{w'c'}/(u_*c_*)$ from a wind-tunnel experiment on the dispersion of heat from an elevated plant source within a model canopy; x is the fetch of source, u_* the friction velocity and c_* the temperature scale $H/(\rho C_p u_*)$ (H is the source power per unit area and C_p the specific heat of air at constant pressure.) Reproduced from Coppin, Raupach & Legg (1986) with the permission of D. Reidel Publishing Company.

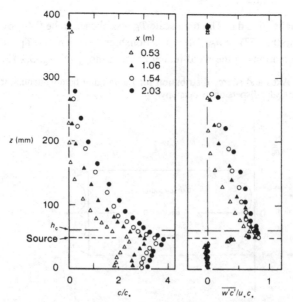

Coppin & Legg, 1986; Coppin, Raupach & Legg, 1986); they represent conditions in the upper part of the canopy at $z = 0.8h_c$. The predictions in Fig. 3.6 are therefore comparable with measurements of scalar (e.g. trace heat) dispersion from an effectively horizontal plane heat source, placed in the same model canopy at a height of $0.8h_c$ (Coppin, Raupach & Legg, 1986). Fig. 3.7 shows how the measured $\bar{c}$ and $\overline{w'c'}$ profiles evolve with increasing x (distance downstream from the leading edge of the heat source).

Examining the predictions in Fig. 3.6, evident features include the sharp 'nose' on the $\bar{c}$ profile at the source height, the rapid equilibration of the $\bar{c}$ profile shape at the source height within the canopy, the approach of the flux profile to zero below the source and a constant flux F_* above the source. All of these correspond well with the wind-tunnel observations in Fig. 3.7. The most significant aspect of Fig. 3.6, however, is the behaviour of K: it is only equal to the far-field value, $K_f = \sigma_w^2 T_L$, well above the canopy, where the travel time for all the dispersing scalar is long enough to satisfy the far-field criteria. Close to and within the canopy the scalar field is a superposition of plumes in or close to the near field. The effect of this can be seen by noting that a plume in the near field is narrower and therefore more concentrated than it would be if far-field diffusion operated everywhere; the actual width is $\sigma_z = \sigma_w t$, whereas the far-field requirement of $\sigma_w(2T_L t)^{1/2}$ is different by the factor $(2T_L/t)^{1/2}$, a large number in the near field. Hence, the narrow, concentrated near-field plumes accentuate the contribution of nearby leaves to the concentration profile, again relative to what would be seen if far-field diffusion operated throughout. This effect causes the $\bar{c}$ profile in Fig. 3.6 to peak sharply at the source height. Because $\bar{c}$ gradients near the source are thereby accentuated, the diffusivity K is diminished relative to K_f, as can also be seen in Fig. 3.6.

We turn now to the phenomenon of counter-gradient fluxes, clearly shown by the

Fig. 3.8. Predicted mean concentration $\bar{c}$ and vertical flux density $\overline{w'c'}$ at $x/h_c = 1000$, for four bimodal source distributions S(z) shown on left. Other conditions as in Fig. 3.7. Reproduced from Raupach (1987) with the permission of the Royal Meteorological Society.

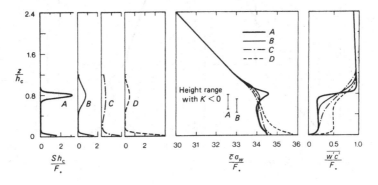

pine forest data in Fig. 3.2. The Lagrangian theory shows that a pronounced counter-gradient flux is possible for a bimodal source distribution, that is, a source density profile $S(z)$ which is concentrated at two heights, one in the foliage and the other at the ground. This is a common distribution of heat or water vapour sources. Fig. 3.8 shows predicted profiles of $\bar{c}$ and $\overline{w'c'}$ for the homogeneous-turbulence canopy, with four bimodal source profiles. Two of these (A and B) show a counter-gradient flux. What happens is that the upper source imparts a bulge to the concentration profile, because of the near-field effect described in connection with Fig. 6, and this can be large enough to make $\partial \bar{c}/\partial z$ positive just below the upper maximum in $S(z)$. However, the presence of the lower source ensures that $\overline{w'c'}$ is always positive, because of scalar conservation. The result is a counter-gradient flux, if the upper source is sufficiently strong and localised as in cases A and B in Fig. 3.8.

Finally, we present a calculation for inhomogeneous turbulence, made using a numerical random-flight model based on the equation recommended by Wilson, Legg & Thomson (1983). Fig. 3.9 shows the assumed wind field, which was based on the model data in Fig. 3.1, and the $\bar{c}$ profile produced by source A in Fig. 3.8. The calculation was stopped at $x/h_c = 120$ to keep the computing time reasonable, but by this stage the $\bar{c}$ profile for inhomogeneous turbulence is very similar to that for homogeneous turbulence, with both profiles showing a region where the $\overline{w'c'}$ flux is counter-gradient. The only significant difference between the two occurs near the

Fig. 3.9. (a): Profiles of $\bar{u}$, σ_W, T_L for inhomogeneous turbulence within and above canopy. (b) and (c): Numerical random-flight calculations of $\bar{c}$ and $\overline{w'c'}$ at $x/h_c = 120$, with source distribution A in Fig. 3.8, made using equation of Wilson, Legg & Thomson (1983) with 10^4 particle flights.

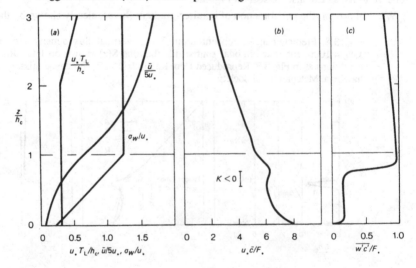

ground, where the decreased σ_W and $\bar{u}$ in the inhomogeneous case causes $\bar{c}$ to be higher than for homogeneous turbulence.

This result, together with our earlier conclusions about the strong influence of the source density $S(z)$ on the concentration field, leads to the suggestion that a rather crude model for the wind field may be adequate to describe scalar transfer from a given source distribution $S(z)$. The most important features to get right in the wind field are the overall magnitudes of T_L and σ_W, as these determine the importance of the near-field effect and the far-field diffusivity $K_f = \sigma_W^2 T_L$.

Conclusions

This chapter has considered the physics of the turbulent transfer of scalars through a canopy. We have made the following points:

(1) All theories for turbulent transfer are based on conservation equations for fluid mass, momentum and mass of scalar. These equations may be treated in an Eulerian (fixed-point) or Lagrangian (fluid-following) framework. Because of the averaging operation necessary to describe the turbulence deterministically, the conservation equations must be augmented with extra information: closure assumptions (such as the gradient-diffusion hypothesis) in the Eulerian theory, and statistical information about fluid particle motions (the transition probability) in the Lagrangian theory.

(2) Turbulence in the canopy is large-scale, the vertical length scale being a substantial fraction of the canopy height h_c. This invalidates the gradient-diffusion hypothesis in the Eulerian framework, and, in the Lagrangian framework, means that both near-field ($t \ll T_L$) and far-field ($t \gg T_L$) dispersion regimes are significant. Here, t is the scalar travel time and T_L the Lagrangian time scale. In the Lagrangian framework, the importance of near-field dispersion arises from the persistence of the turbulent motion, which in turn is related to the dominance of turbulent transfer in a canopy by eddies with vertical length scales comparable with the canopy height.

(3) By noting the different characteristics of near-field dispersion (where $\sigma_Z \sim t$, σ_Z being the vertical width of a plume from an elementary source) and far-field dispersion (where $\sigma_Z \sim t^{1/2}$), one can readily explain the failure of gradient-diffusion theory within the canopy, and observations of counter-gradient fluxes.

(4) A feature of the near-field influence on canopy dispersion is to impress upon concentration profiles, $\bar{c}(z)$, the signature of the source density profile, $S(z)$, more strongly than would be the case with simple gradient diffusion.

(5) Because of the strong influence of $S(z)$ and the relatively weaker influence of the fine details of the inhomogeneous wind field, a rather simple model for the inhomogeneous wind field in the canopy is probably all that is necessary

to calculate $\bar{c}(z)$ quite accurately. However, it is important to have the correct relationships between the overall magnitudes of $\bar{u}$, T_L, σ_W and h_c. Present indications are that T_L is roughly constant at $0.3h_c/u_*$ (where u_* is the friction velocity) within and just above the canopy.

References

Arnold, L. (1974). *Stochastic differential equations: theory and applications.* New York: Wiley-Interscience.

Batchelor, G.K. (1949). Diffusion in a field of homogeneous turbulence: I. Eulerian analysis. *Australian Journal of Scientific Research*, 2, 437–50.

Brown, K.W. & Covey, W. (1966). The energy-budget evaluation of the micrometeorological transfer processes within a cornfield. *Agricultural Meteorology*, 3, 73–96.

Coppin, P.A. (1985). Heat and mass transfer mechanisms above and within plant canopies. *Proceedings of the Third Australasian Conference on Heat and Mass Transfer*, University of Melbourne, 1985, pp. 465–72.

Coppin, P.A., Raupach, M.R. & Legg, B.J. (1986). Experiments on scalar dispersion within a plant canopy, part II: An elevated plane source. *Boundary-Layer Meteorology*, 35, 167–91.

Corrsin, S. (1974). Limitations of gradient transport models in random walks and in turbulence. *Advances in Geophysics*, 18A, 25–60.

Csanady, G.T. (1973). *Turbulent diffusion in the environment.* Dordrecht: D. Reidel.

Deardorff, J.W. (1978). Closure of second- and third-moment rate equations for diffusion in homogeneous turbulence. *Physics of Fluids*, 21, 525–30.

Denmead, O.T. & Bradley, E.F. (1985). Flux-gradient relationships in a forest canopy. In *The Forest-Atmosphere Interaction*, eds. B.A. Hutchison & B.B. Hicks, pp. 421–42. Dordrecht: D. Reidel.

Donaldson, G. duP. (1973). Construction of a dynamic model of the production of atmospheric turbulence and the dispersal of atmospheric pollutants. In *Workshop on Micrometeorology*, ed. D.A. Haugen, pp. 313–92. Boston: American Meteorological Society.

Durbin, P.A. (1983). *Stochastic differential equations and turbulent dispersion.* NASA Reference Publication 1103.

Finnigan, J.J. (1979a). Turbulence in waving wheat. I. Mean statistics and Honami. *Boundary-Layer Meteorology*, 16, 181–211.

Finnigan, J.J. (1979b). Turbulence in waving wheat. II. Structure of momentum transfer. *Boundary-Layer Meteorology*, 16, 213–36.

Finnigan, J.J. (1985). Turbulent transport in flexible plant canopies. In *The Forest-Atmosphere Interaction*, eds. B.A. Hutchison & B.B. Hicks, pp. 443–80. Dordrecht: D. Reidel.

Finnigan, J.J. & Raupach, M.R. (1987). Transfer processes in plant canopies in relation to stomatal characteristics. In *Stomatal Function*, eds. E. Zeiger, G.D. Farquhar & I.R. Cowan, pp. 385–429. Stanford: Stanford University Press.

Lamb, R.G. (1980). Mathematical principles of diffusion modelling. In *Atmospheric Boundary-Layer Physics*, ed. A. Longhetto, pp. 173–210. Amsterdam: Elsevier.

Landau, L.D. & Lifshitz, E.M. (1959). *Fluid Mechanics.* Vol. 6, Course of theoretical physics. English edn, Pergamon Press.

Launder, B.E. (1976). Heat and mass transport. Chapter 6 in *Turbulence*, ed. P. Bradshaw. Berlin: Springer–Verlag.

Legg, B.J. & Raupach, M.R. (1982). Markov-chain simulation of particle dispersion in inhomogeneous flows: the mean drift velocity induced by a gradient in Eulerian velocity variance. *Boundary-Layer Meteorology*, **24**, 3–13.

Legg, B.J., Raupach, M.R. & Coppin, P.A. (1986). Experiments on scalar dispersion within a plant canopy, part III: An elevated line source. *Boundary-Layer Meteorology*, **35**, 277–302.

Lumley, J.L. (1978). Computational modeling of turbulent flows. *Advances in Applied Mechanics*, **18**, 124–76.

Monin, A.S. & Yaglom, A.M. (1971). *Statistical Fluid Mechanics: Mechanics of Turbulence*, Vol. 1. Cambridge, Mass: MIT Press. (Eng. Trans: J.L. Lumley, ed.)

Monteith, J.L. (1973). *Principles of Environmental Physics*. London: Arnold.

Raupach, M.R. (1987). A Lagrangian analysis of scalar transfer in vegetation canopies. *Quarterly Journal of the Royal Meteorological Society*, **113**, 107–20.

Raupach, M.R., Coppin, P.A. & Legg, B.J. (1986). Experiments on scalar dispersion within a plant canopy, part I: the turbulence structure. *Boundary-Layer Meteorology*, **35**, 21–52.

Raupach, M.R., Thom, A.S. & Edwards, I. (1980). A wind tunnel study of turbulent flow close to regularly arrayed rough surfaces. *Boundary-Layer Meteorology*, **18**, 373–97.

Sawford, B.L. (1984). The basis for, and some limitations of, the Langevin equation in atmospheric relative dispersion modelling. *Atmospheric Environment*, **18**, 2405–11.

Taylor, G.I. (1921). Diffusion by continuous movements. *Proceedings of the London Mathematical Society*, **A20**, 196–211.

Tennekes, H. & Lumley, J.L. (1972). *A first course in turbulence*. Cambridge, Mass: MIT Press.

Thom, A.S. (1971). Momentum absorption by vegetation. *Quarterly Journal of the Royal Meteorological Society*, **97**, 414–28.

Thomson, D.J. (1984). Random walk modelling of diffusion in inhomogeneous turbulence. *Quarterly Journal of the Royal Meteorological Society*, **110**, 1107–20.

van Dop, H., Nieuwstadt, F.T.M. & Hunt, J.C.R. (1985). Random walk models for particle displacements in inhomogeneous unsteady turbulent flows. *Physics of Fluids*, **28**, 1639–53.

van Kampen, N.G. (1981). *Stochastic Processes in Physics and Chemistry*. Amsterdam: North-Holland.

Wilson, J.D., Thurtell, G.W. & Kidd, G.E. (1981a). Numerical simulation of particle trajectories in inhomogeneous turbulence. I: systems with constant turbulent velocity scale. *Boundary-Layer Meteorology*, **21**, 295–313.

Wilson, J.D., Thurtell, G.W. & Kidd, G.E. (1981b). Numerical simulation of particle trajectories in inhomogeneous turbulence. II: systems with variable turbulent velocity scale. *Boundary-Layer Meteorology*, **21**, 423–41.

Wilson, J.D., Ward, D.P., Thurtell, G.W. & Kidd, G.E. (1982). Statistics of atmospheric turbulence within and above a corn canopy. *Boundary-Layer Meteorology*, **24**, 495–519.

Wilson, J.D., Legg, B.J. & Thomson, D. (1983). Correct calculation of particle trajectories in the presence of a gradient in turbulent velocity variance. *Boundary-Layer Meteorology*, **27**, 163–9.

Wilson, N.R. & Shaw, R.H. (1977). A higher-order closure model for canopy flow. *Journal of Applied Meteorology*, **16**, 1198–205.

Wong, S.C., Cowan, I.R. & Farquhar, G.D. (1979). Stomatal conductance correlates with photosynthetic capacity. *Nature*, **282**, 424–6.

K.G. McNAUGHTON

4. Regional interactions between canopies and the atmosphere

Introduction

Plant canopies modify their own microclimate. The heat and vapour released into the atmosphere at plant surfaces changes the temperature and humidity of the air in contact with those surfaces. These changes in temperature and humidity, in their turn, modulate the fluxes of heat and vapour from the vegetation. The importance of this 'atmospheric feedback' depends, amongst other things, on the area of the plant canopy (Jarvis & McNaughton, 1986). Small areas of vegetation modify shallow layers of the atmosphere, and local changes in microclimate are small. The influence of a single field extends upwards for perhaps 10 metres. The gradients of temperature and humidity through this layer have been studied in detail by canopy meteorologists.

If a uniform canopy covers an area of some hundreds of square kilometres then the effect of the vegetation will be felt throughout the whole of the turbulent planetary boundary layer, up to a kilometre or so above the ground. On this regional scale, processes affecting the surface energy balance have received very little scientific attention. This situation is now changing under pressure from hydrologists, who want methods for estimating regional evaporation, and climatologists, who must model the surface energy balance to improve predictions from their models of the global circulation of the atmosphere.

The purpose of this chapter is to review efforts to extend canopy energy balance models to the regional scale. First is a brief descriptive account of atmospheric transport processes in the whole planetary boundary layer, to set the scene. An account of some simple energy balance models follows. These models all deal with a convective boundary layer (CBL) capped by an inversion, so they do not cover all atmospheric conditions. However, they do describe conditions under which a substantial part of evaporation from land surfaces takes place. These include daytime anti-cyclonic conditions at mid latitudes and conditions beneath trade-wind inversions at lower latitudes.

Overview of the convective boundary layer

The following paragraphs give a brief description of the structure of a convective boundary layer (CBL). Fuller accounts, with reference to the primary literature, are to be found in the reviews by Webb (1984), Wyngaard (1983) and Caughey (1982).

The CBL is a turbulent layer of the atmosphere that develops from the ground upwards during the daytime. Its depth varies diurnally, from a few hundred metres at dawn to perhaps 1500 m by mid afternoon, so daily growth is a major characteristic. Above the CBL the flow is smoother, and the temperature and humidity profiles are set by atmospheric processes at the scale of the global general circulation.

Sensible heat is released at the ground during the greater part of the day, and this warms the CBL and gives rise to convective motions throughout its depth. Fig. 4.1 is a representation of these convective motions, based on the description by Webb (1977, 1984). Walls of rising warm air develop some tens of metres above the ground, forming a polygonal pattern across the landscape in light winds. These polygons are perhaps 2 km across, depending on height of the CBL, with walls about 100 m thick and rising to a height of 300 m or so. Columnar plumes rise from the vertices of the polygons, and the warmed air continues upward until stopped by opposing buoyancy forces in the inversion that caps the CBL. Sometimes the tops of these thermals are marked by cumulus clouds, which can then be seen scattered across the sky. In stronger winds, and over rougher vegetation, the cross-walls of the polygons become weaker and finally disappear, leaving the thermal updrafts stretched out in rows aligned with the wind. When cumulus clouds occur in these conditions they form rows or 'cloud streets'.

Fig. 4.1. A sketch of convection patterns within a convective planetary boundary layer, when wind speed in the mixed layer is small (Adapted from Webb, 1977).

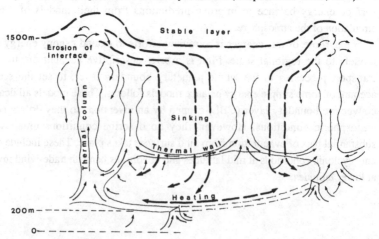

The heated air within thermal updrafts rises quickly, at about 1 m s⁻¹, while the cooler air between them subsides, more slowly but over a larger area. The overall effect is that the air within the CBL becomes quite well mixed from a little way above the ground to the full CBL height. Convection is less effective near the ground, so sizeable gradients of temperature and humidity may occur there. Before further analysis is carried out, two terms need to be defined. Potential temperature (θ) is the temperature that a parcel of air would have if brought adiabatically to a standard pressure, usually 100 kPa. It is given by $\theta = T + \Gamma z$, where Γ is the adiabatic lapse rate (0.01 K m⁻¹). Its gradient is a measure of the stability of air of constant composition. Virtual potential temperature is calculated from $\theta_v = \theta(1 + 0.61q)$, where q is specific humidity in g kg⁻¹. Its gradient is also a measure of the stability of the air, but with a small adjustment to correct for differences in water content. Brutsaert (1982) gives further details.

The effects of convection in the CBL are exemplified in Fig. 4.2, which shows profiles of virtual potential temperature (θ_v) and specific humidity (q) measured over a well-watered landscape. The profiles in Fig. 4.2 are nearly vertical through most of the CBL because of the efficient mixing. Pronounced gradients occur only near the ground and within the capping inversion. The air in the CBL shown is unstable from the ground up to perhaps 100 m, neutral from there to almost 2000 m, and stable at the top in the capping inversion.

The upper part of the CBL is a complex region where mixing occurs between the air within the CBL and that of the stable atmosphere above. The unevenness of the interface is illustrated in Fig. 4.3. In this region rising plumes impinge on the

Fig. 4.2. Soundings of virtual potential temperature, θ_v, and specific humidity, q, measured at Davis, California during a wet spring when the region was uniformly well-watered (from Myrup, Johnson & Pruitt, 1985). Also shown are idealisations of these profiles, as used in simple mixed-layer models. In these models the capping inversion is idealised as a sharp step at height, h, below which θ and q are taken to have constant values, θ_m and q_m, through all of the CBL above the surface layer.

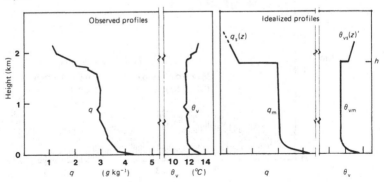

inversion base and cause intermittent bulges in it. Plumes that penetrate into the stable atmosphere above the inversion base are cooler than the surrounding air of the inversion, so buoyancy forces oppose their continued upward motion. The rising air slows and falls back into the CBL, mixing with air from the inversion as it does so. The net effect is that air from the stable layer is carried down through the inversion base and incorporated into the CBL. The process is known as entrainment.

Entrainment cannot proceed indefinitely. As the inversion base is eroded the temperature differential between air in the CBL and the air immediately above it, $\Delta\theta_v$, increases. This increases the buoyancy forces opposing further downward transport across the inversion base. Growth of the CBL must soon cease unless there is also 'warming' from below to reduce $\Delta\theta_v$. Such warming must occur since CBLs have, by definition, upward virtual heat fluxes at their bases. H_v is the virtual heat flux given by:

$$H_v = H + 0.07\lambda E .$$ (1)

H_v is sometimes referred to as the buoyancy flux. Its dominant component is the sensible heat flux, H, with the contribution of the water vapour flux, E, arising because moist air is somewhat less dense than dry air at the same pressure; λ is the latent heat of vaporisation of water; λE is usually the minor component of H_v, except in the humid tropics. During the course of a day it is this 'warming out' process that exerts the predominant control over CBL growth.

Fig. 4.3. Observations of a CBL made using Lidar. The interface between the CBL and the free atmosphere above is made lumpy by the action of convective plumes. Lidar is a laser scanning technique that detects particles or droplets in the atmosphere, so the diagram shows trapping of aerosols within the CBL. A thin cloud lies just above the CBL (Redrawn from Boers, Eloranta & Coulter, 1984).

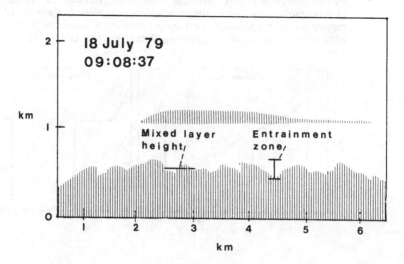

Air incorporated into the CBL from above is both warmer and drier than the air within, tending to make the upper part of the CBL also warmer and drier. The CBL is also warmed from below, so convective overturning leads to very nearly vertical profiles in θ_v, as illustrated in Fig. 4.2. On the other hand, the CBL is moistened at the bottom by addition of evaporated water while it is dried at the top by entrainment. Convective mixing is not quite so successful in eliminating humidity gradients, though the gradients in the main part of the CBL usually are not large, as shown also in Fig. 4.2.

In simple models the CBL is subdivided into two layers: a surface layer comprising perhaps the lowest ten percent, and a mixed layer comprising the remainder. Large gradients of temperature and humidity characterise the surface layer while these gradients are taken as negligible in the mixed layer. The top of the CBL is idealised as a sharp boundary where a sudden change in temperature and humidity may occur, as shown in Fig. 4.2. This simplified scheme is used in the energy balance models discussed here.

We begin our discussion of these simple CBL models with the part most familiar to canopy micro-meteorologists: the surface layer.

Surface layer relationships

Evaporation from the surface vegetation is described by the Penman big-leaf model, which can be written as

$$E = \rho g_c[q^*(T_0) - q_0] , \qquad (2)$$

where ρ is the density of air, g_c is the canopy conductance, T is temperature, q is the specific humidity, and $q^*(T)$ is the saturation specific humidity at temperature T. The subscript '0', here and elsewhere, indicates a value at the canopy 'surface'. Real canopies are rather more complex than is allowed for in the big-leaf model, so the canopy 'surface' is difficult to define exactly. However, a more complex model is not warranted since greater uncertainties arise in the averaging of properties over all vegetation of the landscape. Here we accept that eqn (2) is a useful representation when the vegetation is dry (no intercepted water present) and evaporation from the soil is negligible.

First we make a small change in eqn (2), and write it as

$$E = \rho g_c D_0 , \qquad (3)$$

where D_0 is the surface value of the potential saturation deficit. The potential saturation deficit, D, is the saturation deficit that a parcel of air would have if brought down adiabatically from its own height to the ground. It is given by

$$D = q^*(\bar{\theta}_0) + s(\theta - \bar{\theta}_0) - q . \qquad (4)$$

For our purposes the reference level, $z = 0$, is set at the surface, rather than at the conventional 100 kPa pressure level commonly used by meteorologists. With the ground as reference we have $T_0 = \theta_0$, so $\bar{\theta}_0$ in eqn (4) is equivalent to T_0. The slope of the saturation specific humidity vs temperature relationship, $s = dq^*/dT$, is calculated at mean surface temperature ($\bar{T}_0$). At the ground, where $\theta = T$, D is a linear approximation to the true saturation deficit. Thus eqn (3) is a close approximation to eqn (2).

The potential saturation deficit in the mixed layer (D_m) is an important prognostic variable in CBL models. However, the canopy model, eqn (3) is written in terms of D_0, so we eliminate D_0 following the classic procedure of Penman & Monteith (Monteith, 1973). The surface layer links the ground vegetation to the mixed layer. It can be regarded as 'thin', so fluxes of heat and vapour from the surface can be regarded as passing through without accumulation. Differences of θ and q across the surface layer can then be expressed in terms of the fluxes H and E by the transport equations

$$H = \rho c_p g_{as}(\theta_0 - \theta_m) , \tag{5}$$

and

$$E = \rho g_{as}(q_0 - q_m) , \tag{6}$$

where g_{as} is the aerodynamic conductance for heat and vapour across the whole surface layer, from the canopy up to some level where gradients are no longer significant, and c_p is the specific heat of air. The subscript 'm' indicates values in the mixed layer. We also have the surface energy balance equation

$$H + \lambda E = R_n - G , \tag{7}$$

where R_n is the net radiation into the surface and G is the flux of heat conducted into the ground. This gives all the ingredients needed to derive the Penman–Monteith equation for E, which we write as

$$E = \frac{\varepsilon(R_n - G)/\lambda + \rho g_{as} D_m}{1 + \varepsilon + g_{as}/g_c} , \tag{8}$$

where $\varepsilon = s\lambda/c_p$. This equation differs slightly from the usual Penman–Monteith form in that D_m is the potential rather than the true saturation deficit, and ε is evaluated at the mean surface temperature rather than at a temperature midway between T_0 and T_m. The surface-layer conductance, g_{as}, is measured across the whole surface layer.

We require a method for calculating g_{as}. The gradients of θ and q are largest near the ground and diminish with height towards the top of the surface layer, so most attention should be paid to the profiles of θ and q near the ground. The gradients are small and intermittent at height $|L|$ (Webb, 1984), so this height can be used to define the height of the surface layer. Here L is the Obukhov length scale, given by

$$L = -\frac{\rho c_p T u_*^3}{kg H_v} \, , \qquad (9)$$

where k (= 0.4) is the von Kármán constant and g (= 9.8 m s^{-2}) is the acceleration due to gravity, T is ambient temperature, u_* is the frictional velocity and H_v is the virtual heat flux given by eqn (1).

The aerodynamic conductance of the surface layer, g_{as}, can be calculated from the empirical equations that describe temperature and humidity profiles in the diabatic surface layer. Wyngaard, Arya & Coté (1974) used expressions for θ profiles obtained by Businger *et al.* (1971) and integrated from the surface, which they took to be at the roughness length, $z = z_0$, up to $10|L|$. McNaughton & Spriggs (1986) followed a similar procedure, but used profile forms recommended by Dyer (1974). They took the integration height to be $|L|$, as indicated by Webb (1984), and the surface to be at z_T rather than at z_0, where $\ln(z_0/z_T) = 2$. This approximate correction was proposed by Garratt (1978) to allow for the less efficient transport of heat from canopies compared with that of momentum. The resulting equation for g_{as} is

$$g_{as} = k u_* / \ln(|L|/z_0) \, . \qquad (10)$$

The exact integration height is unimportant since the gradients are small at heights comparable to $|L|$.

Above canopies the profiles of temperature, humidity and trace gases, such as CO_2 and non-reacting pollutants, take the same form in unstable conditions, so eqn (10) should apply equally to transfer of any scalar across the surface layer. Since Obukhov similarity is assumed, eqn (10) requires that $|L| \gg z_0$ and $|L| \ll h$, so it is unlikely to be accurate early in the morning, when h is small and $|L|$ is large, or over forests where z_0 is large.

An alternative approach to estimating g_{as} is to match surface-layer profiles to outer-layer profiles in an assumed overlap region using CBL similarity theory. Brutsaert (1982, 1986) reviews these methods and results. Brutsaert documents some evidence that g_{as} is different for heat and vapour, but no convincing evidence for a difference between g_{as} for heat and vapour was found by Myrup *et al.* (1985). This method gives similar results to profile integration.

Eqn (8) is used as a lower boundary condition to the CBL models discussed here. We now look at the simplest of these.

CBL without entrainment

The earliest models of regional energy balances were developed by McNaughton (1976*a,b*) and Perrier (1980). Both ignored entrainment and treated the CBL as a simple mixed layer with an impermeable lid. McNaughton considered a two-dimensional model with steady net radiation. A 'regional' solution was obtained as the limit at very large distance where advective effects vanish. Perrier examined a one-dimensional model of a CBL in which the net radiation input varied sinusoidally.

Here, as a first step, we examine a simple one-dimensional model of a CBL with steady input of available energy, $(R_n - G)$.

Consider the CBL to be a well-mixed layer of air, of fixed height h, overlying a uniform landscape of transpiring vegetation. The surface layer is regarded as 'thin', so the Penman–Monteith equation, eqn (8), is used to define conditions at the lower boundary of the CBL.

Because heat and vapour are both added, both the temperature and the humidity of the mixed layer must rise continuously. The rate of increase of temperature is given by

$$\rho c_p h \frac{d\theta_m}{dt} = H ,$$
(11)

while the rate of increase of humidity is given by

$$\rho h \frac{dq_m}{dt} = E ,$$
(12)

where h here is held constant. No special allowance has been made for changes in storage of heat and vapour in the surface layer. Surface layer thickness is included within h, so heat and vapour storage changes within it are accounted as if its temperature and humidity were θ_m and q_m. Only storage changes associated with the changes in the shapes of the θ and q profiles in the surface layer have been truly neglected.

Noting that $dD = s d\theta - dq$, we multiply eqn (11) by s, multiply eqn (12) by c_p, and subtract to obtain:

$$\rho c_p h \frac{dD_m}{dt} = sH - c_p E .$$
(13)

H can be substituted in this equation using the energy balance, eqn (7), and then E can be eliminated using the Penman–Monteith equation, eqn (8), to yield a first-order differential equation in the single variable D_m. This can be written as

$$\frac{dD_m}{dt} + \frac{D_m}{\tau} - \frac{D_{eq}}{\tau} = 0 ,$$
(14)

where τ is the time constant, defined below. When $(R_n - G)$ is held constant this equation has the solution

$$D_m = D_{eq} + (D_i - D_{eq}) \exp(-t/\tau) ,$$
(15)

where D_i is the initial value of D_m at $t = 0$, and where

$$D_{eq} = \frac{\varepsilon (R_n - G)}{\rho \lambda g_c (1 + \varepsilon)}$$
(16)

is the final value of D_m that is approached exponentially with the time constant, τ, given by

$$\tau = h \left[\frac{1}{g_{as}} + \frac{1}{(1 + \varepsilon) g_c} \right] .$$
(17)

Substituting D_{eq} for D_m in eqn (8) gives the final evaporation rate as

$$E = \frac{\varepsilon}{(\varepsilon + 1)} \frac{(R_n - G)}{\lambda} , \tag{18}$$

which is known as the 'equilibrium evaporation rate', E_{eq}. This name was first used by Slatyer & McIlroy (1961), who argued that the atmosphere above an extensive wet surface would eventually become saturated, with the resulting evaporation rate given by eqn (18). Later, Priestley & Taylor (1972) noted that their argument was incomplete since they had failed to identify the necessary conditions at the top of the CBL.

Perrier (1980) went beyond this steady-state analysis, and found a solution to eqn (14), where $(R_n - G)$ varies diurnally as a sine wave. His solution also showed that mean daily evaporation approached the mean equilibrium rate in these conditions, while D_m varied sinusoidally about the mean D_{eq}.

This simple model of a closed CBL provides a valuable reference case, but it does not describe real evaporation rates very well. Measurements of evaporation from well-watered agricultural crops and pastures usually show that daytime evaporation exceeds the equilibrium evaporation rate by 20% or more. This is true even where the crops appear to be quite typical of the vegetation of the surrounding region and there are no obvious sources of dry air for advection. The evaporation formulae of Penman (1948) and Priestley & Taylor (1972), which always give evaporation estimates higher than E_{eq}, give reasonably accurate values in these conditions. On the other hand, measurements of evaporation from extensive forests often show evaporation rates lower than the equilibrium rate (e.g. Stewart & Thom, 1973; Shuttleworth *et al.*, 1984). A more realistic model is needed.

CBL with entrainment

In real CBLs, the inversion cap is not impermeable. It allows air to pass downwards through it, so the height of the inversion base rises during the day as air is entrained into the CBL from above. The temperature and humidity of this entrained air is generally warmer and drier than that of the mixed layer, so entrainment tends to raise D_m and increases the evaporation rate.

But this increase in evaporation rate is not without limit. However fast the CBL entrains air from above, air within the mixed layer can become no warmer or drier than the air above, so there is a natural upper limit on D_m, and therefore on the evaporation rate. Another limit is imposed by the need for an upward buoyancy flux to sustain entrainment in the presence of opposing buoyancy forces within the capping inversion. The latent heat flux, λE, cannot exceed available energy, $R_n - G$, by any significant amount since, if it were to do so H_v would vanish, the CBL would stop entraining the drier air, and E would have to decline towards E_{eq}, as shown in the previous section. These limits should arise naturally when we include entrainment in our model for evaporation into a growing CBL.

Let the potential temperature and humidity profiles above the capping inversion be $\theta_s(z)$ and $q_s(z)$, where the subscript 's' indicates values set by synoptic-scale processes, and consider now the incorporation of a thin layer of air, of thickness Δh and potential temperature $\theta_s(h)$, into a mixed layer with potential temperature θ_m and height h. In a time interval Δt the potential temperature in the mixed layer is increased by an amount $\Delta\theta_m$, to $\theta_m + \Delta\theta_m$, given by

$$\theta_m + \Delta\theta_m = \frac{\theta_m h + \theta_s \Delta h}{h + \Delta h} , \tag{19}$$

from which the rate of temperature increase as a result of entrainment is found to be

$$\frac{\Delta\theta_m}{\Delta t} = \frac{(\theta_s - \theta_m)}{(h + \Delta h)} \frac{\Delta h}{\Delta t} . \tag{20}$$

This should be included in the sensible heat budget (eqn (11)), which becomes

$$\rho c_p h \frac{d\theta_m}{dt} = H + \rho c_p(\theta_s - \theta_m) \frac{dh}{dt} . \tag{21}$$

A similar argument leads to addition of an extra term to the humidity budget (eqn (12)), which becomes

$$\rho h \frac{dq_m}{dt} = E + \rho(q_s - q_m) \frac{dh}{dt} . \tag{22}$$

Notice that $\theta_s > \theta_m$ so entrainment warms the CBL, and that $q_s < q_m$ so entrainment reduces its humidity. Both raise D_m and so increase the evaporation rate, via eqn (6). Eqns (21) and (22) can be combined, as before, to give a single equation for D_m, but now there is no simple analytic solution, since h varies with time in an unknown way.

Entrainment was first included in a model for regional evaporation by De Bruin (1983), though eqns (21) and (22) were not given explicitly. De Bruin used a parameterisation developed by Tennekes (1973), who argued that the downward flux of sensible heat entrained through the inversion base should be proportional to the upward sensible heat flux, H, at the ground. Tennekes wrote the right-hand side of eqn (21) as aH, where a was found to be 1.2 from observational data. De Bruin (1983) proposed a similar proportionality between the entrained moisture flux and the surface evaporation rate, writing the right-hand side of eqn (22) as bE. There are no dynamical arguments to support this assumption, but De Bruin found his daily energy balance calculations were insensitive to the chosen value of b. With these simplifications h appeared as an unknown variable only on the left-hand side of eqns (21) and (22). De Bruin then solved these equations using a relationship for height growth of the CBL, $h(t)$, provided by Driedonks (1982).

De Bruin's solution for constant input of $(R_n - G)$ gives the final evaporation rate as

$$E = a\varepsilon(R_n - G)/(a\varepsilon + b) , \tag{23}$$

but the approach to this limit is not a simple exponential, as it was for the closed CBL model, since now the CBL height varies with time. The time 'constant' for the approach to this limit is

$$\tau = \left[\frac{(1 + \varepsilon)}{g_{as}} + \frac{1}{g_c}\right] \frac{h(t)}{(a\varepsilon + b)} . \tag{24}$$

De Bruin observed that this response is slow. With $h \approx 500$ m, $\varepsilon = 2$ (at 18 °C), $a = 1.2$, $b = 0$, and $g_{as} = g_c = 0.02$ m s^{-1}, eqn (24) gives τ more than 7 hours, so it takes almost a day to approach the limit, eqn (23), closely. As the solar radiation input varies through the day, D_m, and so the real evaporation rate, lags behind and falls below the rate given by eqn (23). The shortfall depends on τ, and thus on g_c and g_{as}.

Calculations by de Bruin (1983), using measured $(R_n - G)$ as input, show a dependence of evaporation on canopy conductance. Some results from his calculations, with b set equal to zero, are shown in Fig. 4.4. The lower the value of g_c, the slower the adjustment and so the lower the evaporation rate.

McNaughton & Spriggs (1986) took a different approach to eqns (21) and (22). Rather than parameterise the entrained moisture flux in terms of the surface flux, they sought a simple way to describe the rate of growth of the CBL, dh/dt.

Many schemes have been developed for calculating rate of growth of CBLs. The simplest of these describe only the 'warming out' of the inversion, with no energetic

Fig. 4.4. Midday values of $\lambda E/(R_n - G)$ plotted against canopy resistance, $r_c = 1/g_c$, from calculations by de Bruin (1983, Fig. 2) using observed radiation as input, $b = 0$, and $g_{as} = 20$ mm s^{-1} (———). In replotting de Bruin's results $\varepsilon = 2$ has been assumed, corresponding to observed initial value of θ_{mo} of 18.5 °C. An empirical relationship between daily $\lambda E/R_n$ and r_c, found by Monteith (1965), is plotted for comparison (------). The data points shown are Monteith's. The similarity between these two lines suggests that relationship found by Monteith may result from regional-scale CBL processes.

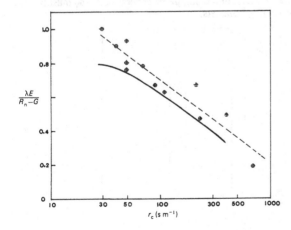

entrainment processes at the inversion base, and so no temperature step, $\Delta\theta_v$. This is illustrated in Fig. 4.5. The rate of growth is then given by

$$\frac{dh}{dt} = \frac{H_v}{\rho c_p \gamma_v h}, \tag{25}$$

where γ_v (= $d\theta_v(z)/dz$ at $z = h$) is the strength of the inversion in virtual potential temperature at height h, as found from an early-morning radiosond ascent. This model for CBL growth is often called the 'encroachment' model, because the CBL grows as the temperature rises in much the same way that a lake encroaches onto a sloping shore as the water level rises. However, Deardorff (1983) has criticised this terminology because it does not describe the key process, which is incorporation.

McNaughton & Spriggs (1986) tested two formulations for dh/dt. The first they made as simple as possible, both to increase its practicality by requiring simpler data inputs, and to obtain an uncluttered view of the processes that are really critical for reproducing the observed behaviour of regional energy balances. McNaughton & Spriggs noted that eqn (25) can also be used when the step in θ_v at the top of the CBL is not zero (Fig. 4.5(b)), provided $d(h\Delta\theta_v)/dt = 0$. Thus eqn (25) can be used when conditions shortly after dawn, at the start of a daytime simulation, show $(\theta_s - \theta_m) \neq 0$. The temperature step then diminishes as the CBL grows, as usually observed. Eqn (25) describes the warming out process but not the dynamics of the entrainment processes which control the size of the step in temperature at the inversion base. Often it does not describe growth of a CBL at all well. By comparison, McNaughton & Spriggs used a second, more realistic formulation, based on dynamical arguments and requiring additional wind information (Driedonks, 1982).

Fig. 4.5. Schematic representations of profiles of θ_v during CBL growth. Panel (a) depicts the so-called 'encroachment only' case where there is no temperature step at the inversion base. This scheme leads to eqn (25) in the text. Panel (b) depicts CBL growth in the presence of a temperature step, $\Delta\theta_v$, at the inversion base. This also leads to eqn (25) when $d(h\Delta\theta_v)/dt = 0$ but, in the general case, $\Delta\theta_v$ must be found from a dynamical model of entrainment.

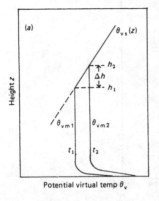
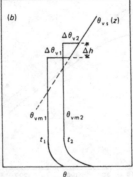

An example of the ability of the McNaughton & Spriggs (1986) model, with dh/dt calculated using eqn (25), to predict the evaporation rate is given in Fig. 4.6. Inputs were observations of $(R_n - G)$, g_{as}, g_c and early-morning profiles of θ and q, from the ground up to the maximum height reached by the CBL during the day. Values of θ_m, q_m and h were calculated by the model. As can be seen, the simulated evaporation rates were quite similar to the observed values. Similar calculations using Driedonks' formulation of entrainment give much better predictions of CBL growth, but similar values for E. Insensitivity of the calculated evaporation rates to the formulation used for entrainment is an encouraging result. It seems that useful models for regional evaporation can be developed without having to deal with difficult problems concerning the dynamics of entrainment.

A useful practical feature of eqn (25) is that it does not depend on observations of windspeed at all. Sensitivity tests by McNaughton & Spriggs showed evaporation calculated by the model to be fairly insensitive to g_{as}, so accurate wind data are not needed at any point for good simulations of the energy balance. The results indicate that regional evaporation is rather insensitive to wind speed.

The model of McNaughton & Spriggs (1986) responds much more rapidly to sudden changes in net radiation or canopy conductance than that of de Bruin (1983). This is because the entrainment terms in eqns (21) and (22) can be large during transients. A sudden increase in R_n, say, leads to a sudden increase in H_v, and so to a sudden increase in dh/dt. While $(\theta_s - \theta_m)$ is limited to a few degrees by the

Fig. 4.6. Comparison of observed and simulated evaporation at Cabauw using the model of McNaughton & Spriggs (1986) with eqn (25) to describe entrainment. Also shown in the observed available energy, which is the principal input to the model.

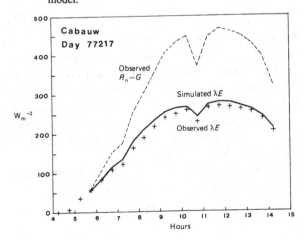

dynamics, $(q_s - q_m)$ may be large. The sudden increase in the influx of drier air into the CBL causes a rapid rise in D_m and a rapid adjustment in E.

Simulation experiments using step changes in net radiation input give model response times of about an hour (McNaughton & Spriggs, unpublished results). In de Bruin's model the rate of adjustment to a sudden change in net radiation is slower by an order of magnitude. This is because the entrained fluxes are made proportional to the surface fluxes, and rapid changes in humidity are not possible. While the two models give similar daily evaporation totals, the internal workings of the models are rather different.

The response time of the CBL is important because it tells us for how long an air mass must pass over uniform terrain before it becomes fully adjusted to the surface. If an air mass moves at 5 m s^{-1} and the CBL adjusts fully in an hour, then the distance travelled by the air mass during adjustment is 18 km. A 'region', then, is a uniform area with a dimension large compared with 18 km – large enough that advection from neighbouring, contrasting regions will have little effect over all.

Notice that this result is for *convective* planetary boundary layers. If an air mass passes from arid land to well–watered land then energy may be consumed in evaporation faster than it is delivered by net radiation, so that the developing boundary layer becomes stable. Models of CBLs have nothing to say about such cases, though we can surmise that advection from the dry upwind area will be felt for much more than 18 km.

Budget for carbon dioxide

We can write a budget equation for the specific concentration of carbon dioxide, c, that is identical in form to eqn (22). Thus

$$\rho h \frac{dc_m}{dt} = A + \rho(c_s - c_m) \frac{dh}{dt} ,$$

(26)

where A is the net CO_2 flux immediately above the canopy in kg m^{-2} s^{-1}; A is taken as positive upwards, in keeping with the convention used for other convective fluxes, so its value is usually negative during the day. Since CO_2 assimilation plays a negligible part in the surface energy balance, this equation can be solved as an independent problem, once $h(t)$ is known from the energy balance calculations described above. Other trace gases, such as SO_2, NO_x, can be treated in a similar way, though in many cases additional source terms would be needed in eqn (26) to account for chemical transformations within the CBL.

We can make an estimate of the size of the drawdown in CO_2 concentration during the convective hours of the day. To do this, rearrange eqn (26) so that it can be written in terms of the derivative of the product $h(c_s - c_m)$, thus

$$\rho d[h(c_s - c_m)]/dt = A ,$$

(27)

then integrate over time, to get

$$(c_s - c_m) = \frac{1}{h}[h_i(c_s - c_{mi}) + \frac{1}{\rho}\int_{t_i}^{t} \theta \, dt'] \quad , \tag{28}$$

where the subscript 'i' indicates an initial value. CO_2 concentrations can be very high at night, sometimes more than 500 ppm by volume, because the CO_2 released by plant respiration and man's fuel burning are often trapped in a shallow, stable, nocturnal boundary layer. After dawn these high concentrations diminish rapidly as photosynthesis begins and heating from the surface fills in, and breaks up the overnight inversion. Within an hour or so the CO_2 concentration within the CBL falls equal to that above the CBL, c_s, which will be close to the mean tropospheric value for that latitude and time of year. We take this time as our starting point, and enquire how much further the CO_2 concentration will be drawn down by regional photosynthesis during the day. That is, we set $(c_s - c_{mi})$ to zero and evaluate only the second term of eqn (28).

If net assimilation of CO_2 for a day is 20 g m^{-2} and the CBL grows to 1000 m then the drawdown in specific CO_2 concentration, c_m, is 17 mg kg^{-1}, or 11 ppm by volume. Higher drawdown could occur in CBLs that grow to lesser heights, either because of a strong overlying inversion or a high evaporation rate that limits H_v to small values. Rates of net assimilation higher than 20 g m^{-2} are possible, but this figure is a regional estimate, including uproductive land. In general terms, the calculation shows that drawdown in c_m will often be of the order of 10 ppm during the convective part of the day.

Observations by Verma & Rosenberg (1976) of CO_2 concentrations at a height of 16 m in an agricultural area in Nebraska, USA are in line with this estimate. They measured high CO_2 concentrations at dawn, but these high values soon disappeared as daytime convection destroyed the overnight ground-based inversion and daily photosynthesis began. Through the convective hours of the summer days, CO_2 concentration decreased by about 10 ppm only to return to higher values as the evening inversion became established. Variations at 2 m, within the surface layer, were larger.

Daytime CO_2 concentration at the canopy 'surface' will be drawn down below that in the mixed layer by an amount $-A/\rho g_{as}$. This expression is obtained from the analogue of eqn (6) for CO_2. With $A = -1$ mg m^{-2} s^{-1} and $g_{as} = 40$ mm s^{-1} this gives a drawdown in surface specific CO_2 concentration of 21 mg kg^{-1} or 13 ppm relative to the mixed layer. These values might characterise a cereal crop during the middle of the day.

Estimating regional evaporation

The discussion above has emphasised meteorological modelling of the CBL. Little has been said of the implications of these larger-scale processes for water use by vegetation. Here we discuss the relationship between regional evaporation and available energy.

The closed CBL model shows that, in the absence of entrainment, evaporation must approach the equilibrium rate, eqn (18). Entrainment can increase E above E_{eq} if $D_m > D_{eq}$, but the process of entrainment itself implies an upper limit for E. Continuing CBL growth can be maintained only if there is an upward virtual heat flux at the ground. If λE approaches 1.07 $(R_n - G)$ then that heat flux vanishes, so $(R_n - G)$ is close to the upper limit for regional evaporation into an inversion capped boundary layer. This must be true no matter how wet the surface and no matter how dry the entrained air. This energy limitation is often taken as self-evident, without need of explanation beyond the energy balance equation, eqn (7), but it does in fact result from the limitation imposed by CBL dynamics on synoptic-scale advection.

If the vegetation has a reasonably high canopy conductance then entrainment raises evaporation above the equilibrium rate. This can be seen by rewriting the (very plastic!) Penman–Monteith equation, eqn (8), in the form

$$E = E_{eq} + \frac{\rho g_{as}(D_m - D_{eq})}{1 + \varepsilon + g_a/g_c}.$$ (29)

Without entrainment $D_m \to D_{eq}$. If $D_s > D_{eq}$ the effect of entrainment will be to raise D_m above D_{eq} and to increase evaporation above the equilibrium rate. On this basis McNaughton *et al.* (1979) proposed that evaporation from unstressed vegetation should always lie in the range $E_{eq} < E < (R_n - G)/\lambda$, and this was used to explain evaporation rates from agricultural crops and pastures, which are commonly observed to he higher than the equilibrium rate.

It is apparent from eqn (16), however, that D_{eq} can be quite large when canopy conductance is small and net radiation is large; D_s will not always be greater than D_{eq}. Forests, in particular, may have low canopy conductance, even in well-watered conditions (Jarvis, James & Landsberg, 1976). The results shown in Fig. 4.4 indicate that the proportion of available energy used in evaporation declines quite significantly as canopy conductance decreases. Small E implies large H, so the CBL will grow more rapidly when canopy conductance is small, and a lot of air will be entrained from above. CBL growth will be particularly rapid if the capping inversion is weak. In this case D_m will approach D_s. The Penman–Monteith equation, eqn (8), with D_s in place of D_m then gives an upper limit to evaporation, and this may be less than E_{eq}. Forests commonly have canopy conductances less than 10 mm s^{-1} and evaporation rates from dry canopies are often less than the equilibrium rate (Jarvis *et al.*, 1976), in accord with these predictions.

Rapidly-growing field crops and pastures with complete ground cover, on the other hand, commonly have canopy conductances in the range 20–50 mm s^{-1} and evaporation rates larger than E_{eq}. Often evaporation from agricultural crops and pastures are broadly in agreement with the Priestley & Taylor equation for estimating large-scale evaporation, $E = 1.26\,E_{eq}$. Penman's specification of 'potential transpiration' 'from a field in the midst of an extended area of similar vegetation' as 'the amount of water transpired per unit time from a short green crop, completely shading the ground, of uniform height and never short of water' (Penman, 1956) seems to fit rather well the situations where we expect E to lie in the range $E_{eq} < E < (R_n - G)/\lambda$, provided that the 'extended area of similar vegetation' is a region large enough for advective effects within the CBL to be negligible.

Like the Priestley & Taylor equation, the Penman (1948) equation always gives estimates of E in the range $E_{eq} < E < (R_n) - G)/\lambda$ when used with data collected in such conditions of 'potential evaporation'. Exceptions may occur in particularly windy conditions, but CBL models show that real regional evaporation is rather insensitive to wind speed so the Penman estimates of E are probably too high in such cases. Both equations therefore give estimates of E that are in the correct range. This might be sufficient to explain their empirical successes; an explanation is needed because the original physical derivations of the two equations are insufficient to explain their successful application to non-wet crops and pastures (Thom & Oliver, 1977).

For forests, where canopy resistances are typically greater than 100 s m^{-1}, the regional evaporation rate can fall below E_{eq} so both equations will often overestimate the real evaporation rate. This explains why the Penman and Priestley & Taylor equations are of so little value for estimating evaporation from forests (Shuttleworth & Calder, 1979).

General comments

It may be observed that the regional energy balance models discussed above are limited in scope to convective boundary layers. They cannot deal with early morning, late afternoon or nighttime hours when ground-based inversions are present, and they cannot deal with disturbed conditions, such as during the passage of fronts. They exclude possible effects of advection within the CBL. Therefore they can have little to say about such important issues as interception losses from wetted forests.

The models are also quite crude and have had limited testing. Perhaps the major achievement so far is simply that they are formulated as true planetary boundary layer models. Evaporation research is now connected to the rest of meteorology via a consistent hierarchy of models. Since a great deal is known about planetary boundary layers, we can expect more of this knowledge to be applied to problems of the surface energy balance quite soon. Results for stable boundary layers should be forthcoming, so that calculations can be extended over entire days. The model of van Ulden & Holstag (1985) already does this, though they use an empirical equation in place of eqn (8). As always, the main difficulty will probably be in assembling sufficient data to use the models effectively.

References
Boers, R., Eloranta, E.W. & Coulter, R.L. (1984). Lidar observations of mixed-layer dynamics: tests of parameterized entrainment models of mixed-layer growth rate. *Journal of Climate and Applied Meteorology*, **23**, 247–66.
Brutsaert, W. (1982). *Evaporation into the Atmosphere*. Dordrecht: Reidel.

Brutsaert, W. (1986). Catchment-scale evaporation and the atmospheric boundary layer. *Water Resources Research*, **22**, 395–455.

Businger, J.A., Wyngaard, J.C., Izumi, Y. & Bradley, E.F. (1971). Flux-profile relationships in the atmospheric surface layer. *Journal of the Atmospheric Sciences*, **28**, 181–9.

De Bruin, H.A.R. (1983). A model for the Priestley–Taylor parameter α. *Journal of Climate and Applied Meteorology*, **22**, 572–8.

Caughey, S.J. (1982). Observed characteristics of the atmospheric boundry layer. In *Atmospheric Turbulence and Air Pollution Modelling*, eds. F.T.M. Nieustadt & H. van Dop, pp. 107–58. Dordrecht: Reidel.

Deardorff, J.M. (1983). Comments on 'The daytime planetary boundary layer: a new interpretation of Wangara data' by P.C. Manins. *Quarterly Journal of the Royal Meteorological Society*, **109**, 677–81.

Driedonks, A.G.M. (1982). Models and observations of the growth of the atmospheric boundary layer. *Boundary-Layer Meteorology*, **33**, 283–306.

Dyer, A.J. (1974). A review of flux-profile relationships. *Boundary-Layer Meteorology*, **7**, 363–72.

Garratt, J.R. (1978). Transfer characteristics for a heterogeneous surface of large aerodynamic roughness. *Quarterly Journal of the Royal Meterological Society*, **104**, 491–502.

Jarvis, P.G., James, G.B. & Landsberg, J.J. (1976). Coniferous forest. In *Vegetation and the Atmosphere. Volume 2*, ed. J.L. Monteith, London: Academic Press, pp. 171–240.

Jarvis, P.G. & McNaughton, K.G. (1986). Stomatal control of transpiration: scaling up from leaf to region. In *Advances in Ecological Research*, **15**, 1–47. London: Academic Press.

McNaughton, K.G. (1976a). Evaporation and advection I: evaporation from extensive homogeneous surfaces. *Quarterly Journal of the Royal Meteorological Society*, **102**, 181–91.

McNaughton, K.G. (1976b). Evaporation and advection II: evaporation downwind of a boundary separating regions having different surface resistances and available energies. *Quarterly Journal of the Royal Meteorological Society*, **10**, 193–202.

McNaughton, K.G., Clothier, B.E. & Kerr, J.P. (1979). Evaporation from land surfaces. In *Physical Hydrology. New Zealand Experience*, eds. D.L. Murray & P. Ackroyd, pp. 97–119. Wellington: New Zealand Hydrological Society.

McNaughton, K.G. & Spriggs, T.W. (1986). A mixed layer model for regional evaporation. *Boundary-Layer Meteorology*, **34**, 243–62.

Monteith, J.L. (1965). Evaporation and Environment. In *The State and Movement of Water in Living Organisms*, ed. G.E. Fogg. Society of Experimental Biology Symposium No. 19, Cambridge University Press. pp. 205–34.

Monteith, J.L. (1973). *Principles of Environmental Physics*. London: Edward Arnold.

Myrup, I.O., Johnson, C.D. & Pruitt, W.O. (1985). Measurement of the atmospheric boundary-layer resistance law for water vapor. *Boundary-Layer Meteorology*, B33, 105–11.

Penman, H.L. (1948). Natural evaporation from open water, bare soil and grass. *Proceedings of the Royal Society, London*, Series A **193**, 120–46.

Penman, H.L. (1956). Evaporation: an introductory survey. *Netherlands Journal of Agricultural Science*, **4**, 9–29.

Perrier, A. (1980). Etude micro-climatique des relations entre les propriétés de surface et les charactéristiques de l'air: application aux échanges régionaux. In *Météorologie & Environnement*. France: EVRY.

Priestley, C.H.B. (1959). *Turbulent Transfer in the Lower Atmosphere.* Chicago: University of Chicago Press.

Priestley, C.H.B. & Taylor, R.J. (1972). On the assessment of surface heat flux and evaporation using large-scale parameters. *Monthly Weather Review,* **100,** 81–92.

Shuttleworth, W.J. & Calder, I.R. (1979). Has the Priestley–Taylor equation any relevance to forest evaporation? *Journal of Applied Meteorology,* **18,** 639–46.

Shuttleworth, W.J. *et al.* (1984). Eddy correlation measurements of energy partition for Amazonian forest. *Quarterly Journal of the Royal Meteorological Society,* **110,** 1143–62.

Slatyer, R.O. & McIlroy, I.C. (1961). *Practical Microclimatology.* Melbourne: C.S.I.R.O.

Stewart, J.B. & Thom, A.S. (1973). Energy budgets in a pine forest. *Quarterly Journal of the Royal Meteorological Society,* **99,** 154–70.

Tennekes, H. (1973). A model for the dynamics of the inversion above a convective boundary layer. *Journal of the Atmospheric Sciences,* **30,** 558–67.

Thom, A.S. & Oliver, H.R. (1977). On Penman's equation for estimating regional evaporation. *Quarterly Journal of the Royal Meteorological Society,* **103,** 345–57.

van Ulden, A.P. & Holstag, A.A.M. (1985). Estimation of atmospheric boundary layer parameters for diffusion applications. *Journal of Climate and Applied Meteorology,* **24,** 1196–1207.

Verma, S.B. & Rosenberg, N.J. (1976). Carbon dioxide concentration and flux in a large agricultural region of the great plains of North America. *Journal of Geophysical Research,* **81,** 399–405.

Webb, E.K. (1977). Convection mechanisms of atmospheric heat transfer from surface to global scales. In *Second Australian Conference on Heat and Mass Transfer. The University of Sydney, February 1977,* ed. R.W. Bilger, pp. 523–39.

Webb, E.K. (1984). Temperature and humidity structure in the lower atmosphere. In *Geodetic Refraction – Effects of Electromagnetic Wave Propagation Through the Atmosphere,* ed. F.K. Brunner, pp. 885–1141. Berlin: Springer.

Wyngaard, J.C. (1983). Lectures on the planetary boundary layer. In *Mesoscale Meteorology – Theories, Observations and Models,* eds. D.K. Lilly & T. Gal-Chen, pp. 603–50. Dordrecht: Reidel.

Wyngaard, J.C., Arya, S.P.S. & Coté, O.R. (1974). Some aspects of structure of convective planetary boundary layers. *Journal of the Atmospheric Sciences,* **31,** 747–54.

H. VAN KEULEN, J. GOUDRIAAN AND
N.G. SELIGMAN

5. Modelling the effects of nitrogen on canopy development and crop growth

Introduction

Green plants utilise the sun's energy to synthesise organic compounds from carbon dioxide and water. The pioneering work, concurrently carried out by Liebig in Germany and Lawes & Gilbert in the UK more than a century ago, conclusively showed that plants must take up inorganic nutrients from the soil to produce these organic components. Since that discovery it has been established that many elements are necessary for optimum functioning of the biochemical machinery of the plant. Most of these are necessary in such small amounts, however, that the supply from the seed, or from natural sources suffices. In agriculture the situation is often different for the macro-elements nitrogen, phosphorus and potassium that are needed in such large quantities, especially where crop management practices aim at very high yields, that the supply from natural sources falls far short of the demand. Fertiliser experiments show that, up to a certain level, addition of these elements from a fertiliser bag leads to higher yields. Unfortunately, interpretation of these fertiliser experiments seldom exceeds the derivation of the optimum nutrient application rate for the conditions of the experiment, either in physical or in economic terms. The lack of explanatory conclusions hinders the use of such results for predictive purposes, for example, in the formulation of fertiliser recommendations for the farmer. This lack of predictive power is especially serious in the case of nitrogen where at present the price of the fertiliser in western economies does not encourage restricted use by the farmer , but where the cost to society, because of air and water pollution due to excessive application of both organic and inorganic manures, may become prohibitive.

A more promising approach would seem to be to describe the effects of nitrogen or its deficiency in terms of the processes that determine crop growth and yield, for example canopy development, and to predict on that basis the nutrient requirements for a certain target yield (Greenwood, 1982). In this chapter we have used an analysis of growth of spring wheat crops as an example of this approach. However, despite more than 100 years of research in the field of plant nutrition, many of the required relationships appear to be either totally absent or, at best, ambiguous. Nevertheless, some results are presented from a model in which these relations have been incorporated in a coherent framework.

The principles underlying the description of the nitrogen economy of the crop would in our view also be applicable for other annual crops. In many instances it appeared that the relationships used were identical for different species. Perennial species behave differently in that recirculation of nitrogen between above- and below-ground plant parts is an important process.

The relationship between nitrogen status and dry matter production

Dry matter production, either the total or that accumulated in a specific plant part such as the tubers or the grains, is the product of the length of the production period and the mean rate of dry matter accumulation during that period. The effect of nitrogen on crop performance can thus be described by accounting for its effect on both components.

Length of the growing period

The phenological development of the plant, i.e. the rate and order of appearance of vegetative and reproductive plant organs, is governed both by genotype, and by environmental factors, notably day length and temperature (van Dobben, 1962a). For at least some cultivars of spring wheat the effects of day length are insignificant and, moreover, for most environments cultivars are available that suit the particular photoperiodic characteristics of that environment. The driving force for development then becomes temperature and the relevant variable is the temperature of the stem apex which can be approximated by either air or canopy temperature.

Nitrogen deficiency may cause stomatal closure at higher plant water potentials (Radin & Ackerson, 1981), or a reduction in water uptake by increased root resistance (Radin & Boyer, 1982) and a consequent reduction in transpiration (Shimshi, 1970a,b; Shimshi & Kafkafi, 1978) which will alter the energy balance of the canopy and hence its temperature. In the field, differences of up to 4 °C in canopy temperatures have been measured between fields optimally supplied with N and fields under nitrogen stress (Seligman, Loomis, Burke & Abshahi, 1983). Under certain climatic conditions an indirect effect of nitrogen shortage on phenological development can thus be expected. Such effects have indeed been reported where a field-grown crop of wheat under nitrogen deficient conditions reached maturity up to 5 days earlier than a crop growing with adequate nitrogen (Seligman et al., 1983). This delay is equivalent to somewhat less than 1°C temperature difference during the main growing period between booting and early grain fill.

Severe stress can of course stop development completely, although it is not clear at what point this can happen. This phenomenon may be the basis of the observation that relatively severe stress can delay phenological development (Angus & Moncur, 1977).

The effect of nitrogen status on the actual transpiration of a wheat canopy is discussed in more detail later in this chapter.

Dry matter production

Carbon dioxide assimilation by the leaf canopy. The influence of nitrogen deficiency in the vegetation on dry matter accumulation, total production, and yield is well documented, but the effects on the basic processes of assimilation and respiration are far less clear. The rate of CO_2 assimilation at different levels of nitrogen concentration in the leaves has been determined for many plant species, such as maize (*Zea mays*) (Goudriaan & van Keulen, 1979; Ryle & Hesketh, 1969), sunflower (*Helianthus annuus*) (Goudriaan & van Keulen, 1979), cotton (*Gossypium hirsutum*) (Wong, 1979; Ryle & Hesketh, 1969), sugar beet (*Beta vulgaris*) (Nevins & Loomis, 1970), rice (*Oryza sativa*) (Cook & Evans, 1983*a,b*; Yoshida & Coronel, 1976; Takeda, 1961), pasture grasses, both those with C_3 and C_4 photosynthetic pathways (Woledge & Pearse, 1985; Bolton & Brown, 1980; Lof, 1976; Wilson, 1975*a,b*), wheat (*Triticum aestivum*) (Marshall, 1978; Osman, Goodman & Cooper, 1977; Dantuma, 1973; Osman & Milthorpe, 1971; Khan & Tsunoda, 1970*a,b*), soya bean (*Glycine max*) (Boon–Long, Egli & Leggett, 1983; Lugg & Sinclair, 1981; Boote, Gallaher, Robertson, Hinson & Hammond, 1978), *Eucalyptus* spp. (Mooney, Ferrar & Slatyer, 1978) and tung (*Aleurites* sp.) (Loustalot, Gilbert & Drosdoff, 1950). In all situations where nitrogen concentration of the leaves was determined, there was a strong correlation between the nitrogen concentration in the leaves and their photosynthetic performance.

Leaf nitrogen concentration can be expressed either on an area basis or on a dry weight basis. In the literature cited above, both methods have been used, but where the specific leaf area is reported, the data can be expressed on a common basis. In Fig. 5.1 some data for C_3 species are summarised from situations where the incident photon flux density during the measurements was high enough to ensure light saturation. The data suggest a linear relation between nitrogen concentration and net CO_2 assimilation rate, at least up to a nitrogen concentration of 0.06 kg kg^{-1}. Some of the residual variability could be due to different ages or development stages of the experimental material. Net assimilation becomes zero at a nitrogen concentration of 3.8 g kg^{-1}.

Net CO_2 assimilation as a function of nitrogen concentration expressed on an area basis is given in Fig. 5.2. Zero assimilation occurs in this case at a nitrogen concentration of 0.2×10^{-4} kg m^{-2}. The two slopes would be identical at a value of the specific leaf area of 31.25 m^2 kg^{-1}, a rather high value, suggesting that most experiments were conducted on young, thin leaves.

The general conclusion from the data presented here is thus that the maximum rate of carbon dioxide assimilation is linearly related to nitrogen concentration over a wide range of concentrations. Sometimes in individual experiments a 'saturation'-type curve seems more appropriate, but that is difficult to derive from the bulked data.

No significant effect of leaf nitrogen concentration on the light use efficiency at low irradiance has been shown (Cook & Evans, 1983*a,b*; Wilson, 1975*a,b*) but it is

Fig. 5.1. The relation between nitrogen concentration in the leaf blade (on a dry weight basis) and the maximum rate of net CO_2 assimilation. ●, ○ *Oryza sativa* (Yoshida & Coronel, 1976); ▼ *Triticum aestivum* (Dantuma, 1973); ▽ *Triticum aestivum* (Marshall, 1978); ⊙ *Panicum* spp. (Brown & Wilson,1983); + *Hordeum murinum,* ⊕ *Phalaris minor* (Lof, 1976); x *Lolium perenne* (Wilson, 1975); □ *Oryza* spp. (Cook & Evans, 1983a,b); ■ *Aleurites* sp. (Loustalot *et al.*, 1950); ▲ *Festuca arundinacea* (Bolton & Brown, 1980). The calculated regression line ($r^2 = 0.77$) has a slope of 22.0 mg CO_2 m^{-2} s^{-1} for each unit increase in nitrogen concentration.

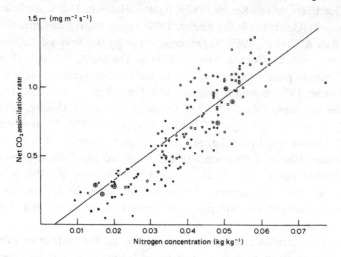

Fig.5.2. The relation between nitrogen concentration in the leaf blade (on an area basis) and the maximum rate of net CO_2 assimilation: x *Beta vulgaris* (Nevins & Loomis, 1970); ●, ○ *Oryza* spp. (Cook & Evans, 1983a,b), Δ *Oryza sativa* (Yoshida & Coronel,1976); + *Glycine max* (Boon–Long *et al.*, 1983).The calculated regression line ($r^2 = 0.76$) has a slope of 7.1 mg CO_2 m^{-2} s^{-1} for each unit increase in nitrogen concentration.

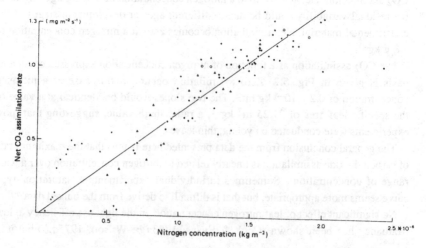

possible that small differences in the slope are responsible for some of the measured differences in assimilation rate between canopies with different leaf nitrogen concentrations. For the time being, the influence of nitrogen concentration on CO_2 assimilation can thus be modelled by relating the value of the light saturated CO_2 assimilation rate of individual leaves, F_m (kg CO_2 ha^{-1} h^{-1}), to the nitrogen concentration in the leaf blades, N (kg kg^{-1}):

$$F_m = 725 N - 2.75 ; F_m > 0 \tag{1}$$

Partitioning of assimilates between respiration and growth of various organs. The assimilates fixed in the photosynthetic process are used in various processes in the plants including maintenance respiration. Sink strength, which is probably related to the number of growing cells in a particular organ (Sunderland, 1960), is an important determinant of assimilate distribution at any time.

Maintenance of the various living plant parts has presumably first priority. Maintenance respiration depends of the weight of an organ, its chemical composition, particularly its nitrogen concentration (reflecting the rebuilding of continuously degrading proteins) and the ambient temperature (Penning de Vries, 1975). In the model, the carbohydrate requirement for maintenance respiration of each organ, i.e. roots, leaves, stems and grains, is calculated first. Although quantitative estimates of maintenance respiration have been proposed, a great deal of uncertainty still exists and in some cases it has been necessary to introduce 'fudge' factors to describe reality sufficiently well (de Wit *et al.*, 1978).

The effect of nitrogen concentration on maintenance respiration is accounted for by a multiplication factor that ranges between 1 and 2, which is about the range found in maintenance requirements per unit dry weight between protein-poor and protein-rich materials (Penning de Vries, 1975). Although quantitative information is rather scarce (Hanson & Hitz, 1983), protein turnover is presumably low at very low nitrogen concentrations, and in any case has a small energy requirement compared to that needed for maintenance of the ionic balance within the cells and the transport of assimilation products.

The assimilates remaining after subtraction of the maintenance requirements of the live organs of the crop, are available for the production of structural plant material and are allocated to five compartments: leaf blades, stems and leaf sheaths, roots, grains and a reserve pool of primary photosynthetic products. When growth conditions are optimum, the proportion of net assimilate allocated to each compartment is a function only of the phenological state of the vegetation, representing the varying sink strengths of the various organs. When growing conditions are sub-optimum the partitioning changes. Whether this is an active process, or the result of a differential influence of stress on the growth of different organs is difficult to judge.

Brouwer (1963; 1965) suggested that the conversion of primary photosynthates into above-ground structural plant material is more inhibited by insufficient moisture supply than is CO_2 assimilation. As a result, the level of reserve carbohydrates in the plant increases and this increases their availability for growth of the root system. As a consequence, water shortage changes the partitioning of assimilates between shoot and root. Brouwer referred to this phenomenon as the 'functional balance'.

Nitrogen shortage in the vegetation also favours the growth of roots at the expense of above-ground material (Cook & Evans, 1983a; Campbell, Davidson & Warder, 1977; Wilson & Haydock, 1971; Colman & Lazenby, 1970; Brouwer, 1965; Brouwer, Jenneskens & Borggreve, 1962; McLean, 1957), which may be the result of the same functional balance. The partitioning between leaf blades and other above-ground organs also changes in nitrogen deficient conditions and generally results in a lower leaf weight ratio (Campbell, Davidson & McCaig, 1983; van Os, 1967; McNeal, Berg & Watson, 1966; Boatwright & Haas, 1961). However, the instantaneous effect of sub-optimum nitrogen concentrations in the tissue on partitioning of assimilates is difficult to quantify from existing experimental data. In the model, nitrogen stress at any particular point in time is defined as the difference between the maximum nitrogen concentration at a certain development stage and the actual nitrogen concentration, expressed as a fraction of the range between the maximum and the minimum nitrogen concentration.

These influences can thus be described schematically by assuming a growth check on the shoot compartments (leaf blades and stem) when nitrogen shortage occurs. The resulting 'surplus' carbohydrate is partitioned between roots and the reserve pool. When stress is alleviated and reserve carbohydrates have accumulated, some can become available for subsequent leaf growth.

Conversion of assimilates into dry matter. The assimilates allocated to the various sinks are a mixture of carbohydrate and nitrogenous compounds. These primary products must be converted into structural plant material, and the energy required for this conversion must be taken into account. The magnitude of this growth respiration also depends on the chemical composition of the material being formed (Penning de Vries, Brunsting & van Laar, 1974; Penning de Vries, 1974). As a first approximation the composition can be defined in terms of protein and carbohydrate only, as these components constitute the major part of the plant material. Since the proteins are assumed to be formed from nitrates only the costs of reduction have to be taken into account.

The rate of increase in plant dry weight is thus obtained by dividing the rate of assimilate supply by the specific assimilate requirement factor. The latter is defined as 1.21 times the fraction of carbohydrates in the currently formed material plus 2.27 times the fraction of proteins (Penning de Vries, 1974). These values, based on detailed biochemical pathway analysis are reasonably well-established, but the uncertainty in this case originates mainly from the assumption that all proteins are formed from nitrate. This description results in higher conversion efficiencies for tissues with a lower nitrogen concentration.

The effect of nitrogen status on transpiration

The actual rate of transpiration of a canopy depends on the potential rate, dictated by meteorological conditions, and on the availability of water in the rooted soil profile.

Many studies have indicated that water use efficiency, i.e. the amount of dry matter produced per unit of water consumed, increases with higher nitrogen availability (van Keulen, 1975; Black, 1966; Viets, 1962). Interpretation of these results is difficult when no distinction has been made between transpiration by plants and evaporation from the soil surface. Plants growing under nitrogen stress generally have a much smaller leaf area than plants growing under optimum nutrient conditions, and so canopy closure occurs much later. Consequently, a larger proportion of water is lost directly from the soil surface, so water use efficiency is reduced. In the early experiments on water use, where direct evaporation from the soil surface was prevented, moderate nitrogen stress had hardly any effect on water use efficiency (Tanner & Sinclair, 1983; de Wit, 1958).

Recent experiments where assimilation and transpiration of plant species were determined on individual leaves with different nitrogen concentrations, have confirmed the latter hypothesis for maize (Goudriaan & van Keulen, 1979; Wong, Cowan & Farquhar, 1979) and *Panicum maximum* (Bolton & Brown, 1980), but not for tall fescue (*Festuca arundinacea*) and *Panicum milioides* (Bolton & Brown, 1980). In the latter species the ratio of apparent photosynthesis to transpiration increased almost twofold over a range of nitrogen concentrations in the leaf from 0.01 to 0.05 $kg \ kg^{-1}$.

A comprehensive study of the interactions between nitrogen and water stress, mainly in cotton, has been conducted by Radin and associates (Radin, 1983; Radin & Boyer, 1982; Radin & Ackerson, 1981; Radin & Parker, 1979a,b). They found that in nitrogen-deficient plants stomatal closure occurs at much higher plant water potentials than in plants adequately supplied with nitrogen, i.e. at values of −1 MPa compared with −1.8 MPa (Radin & Ackerson, 1981). On days of high irradiance the leaf water potential of well-watered wheat plants can remain below −1 MPa for most of the day (Hochman, 1982; Martin & Dougherty, 1975).

Shimshi (1970a,b) has shown that transpiration from nitrogen-deficient plants is reduced at high levels of soil moisture, but that near wilting point the situation is reversed, possibly because the much higher proportion of cell wall constituents in the nitrogen-deficient plants reduces stomatal sensitivity. Consequently, not only is stomatal opening restricted under soil moisture conditions, but full stomatal closure is prevented near wilting point. Evidence for greater stomatal opening with better nitrogen nutrition has also been found in rice (Ishihara, Ebara, Hirawasa & Ogura, 1978; Yoshida & Coronel, 1976), wheat (Shimshi & Kafkafi, 1978), sunflower and maize (Goudriaan & van Keulen, 1979).

Radin & Boyer (1982) have shown that root conductivity is lower in nitrogen-deficient sunflower plants. Lower turgor may occur leading to stomatal closure. Lower transpiration rates due to nitrogen deficiency may either be caused by stomatal closure at higher leaf water potentials or by lower root conductivity. In the former case moisture stress would be secondary and possibly minor compared to nitrogen stress; in the latter case moisture stress induced by nitrogen deficiency would be dominant.

Data on the relation between leaf conductance to water vapour and leaf nitrogen concentration for rice have been published by Yoshida & Coronel (1976). Fig. 5.3 is derived from their data and relates leaf conductance to nitrogen concentration in the leaf. In these data, leaf conductance includes boundary layer conductance, so that stomatal conductance would necessarily be to the left of the regression line. There appears to be a linear relationship between leaf conductance and nitrogen concentration in the leaf. Such a relation suggests stomatal control through the CO_2 concentration in the sub-stomatal cavity (Goudriaan & van Laar, 1978). In that case, stomatal opening is regulated in such a way that the CO_2 concentration either remains constant or has a

Fig. 5.3. The relation between the nitrogen concentration in the leaf blade and total conductance for water vapour exchange for individual rice leaves (Source: Yoshida & Coronel,1976).

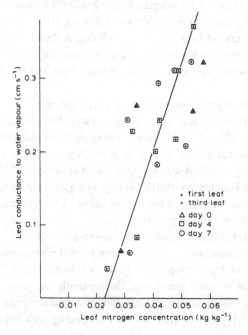

fixed ratio to the external concentration. Any impairment of assimilation will lead then to proportional stomatal closure and decreased transpiration. It is, however, not yet clear under what conditions this phenomenon can be expected in the field.

The effect of nitrogen shortage on water use is therefore difficvlt to predict and must be carefully reconsidered for each situation.

Nitrogen in the plant

Nitrogen uptake

Uptake of nitrogen is dependent on both the demand by the plant and on the availability of nitrogen in the soil. Nitrogen is needed in the plant for synthesis of new tissue, so that as plant weight increases the demand for nitrogen increases. As more structural carbohydrate is accumulated, however, the ratio of nitrogen to total biomass in each of the plant parts falls, even when nitrogen is available in surplus (Vos, 1981; Seligman, van Keulen, Yulzari, Yonathan & Benjamin, 1976; Dilz, 1964; van Burg, 1962; van Dobben, 1962b; 1960). When nitrogen supply is non-limiting, there is a negative linear relationship between the nitrogen concentration in plant organs and the development stage of the crop. The total nitrogen concentration in the leaves falls from an initial value of about 0.06 kg kg^{-1} to 0.02 kg kg^{-1} at maturity. The nitrogen concentration in stem tissue falls from around 0.03 kg kg^{-1} at the onset of stem elongation to about 0.008 kg kg^{-1} at maturity. Maximum nitrogen concentrations in the root are more variable than in the shoot. Appropriate values under conditions of surplus supply from the soil would be about 0.035 kg kg^{-1} at seedling emergence and 0.01 kg kg^{-1} at maturity.

In the model, the nitrogen demand of any plant part at any point in time is defined as the difference between the maximum amount attained under optimum nitrogen supply and the actual amount in the tissue at that moment. The total nitrogen demand of the canopy is then the sum of the nitrogen demands of the component parts or organs, i.e. the leaves, stem and roots. Grain nitrogen is assumed to be supplied by translocation from the vegetative organs and so does not contribute directly to nitrogen demand.

Availability of nitrogen to the vegetation depends both on the amount present in the soil and on the extent and density of the plant's root system (van Keulen, Seligman & Goudriaan, 1975). In the wheat crop, rooting density is generally relatively high, i.e. greater than 1 cm root length cm^{-3} soil (e.g. Gajri & Prihar, 1985; Lupton, Oliver, Ellis, Barnes, Howse, Welbank & Taylor, 1974), and so most of the mineral nitrogen in the rooting zone is available for uptake within one day and virtually all within two days, provided that all or the greater part of that nitrogen is present in the form of nitrate. Consequently, diffusion can supply the balance between demand and mass flow even when mass flow is very low. Early in the season, before the root system of the crop is fully expanded horizontally, some of the soil nitrogen within the rooted depth may be unavailable. This is modelled by defining a maximum uptake rate determined by the extent of the root system. Daily nitrogen uptake is defined as the demand of the crop, the amount available in the soil or the maximum uptake rate, whichever is the highest. This formulation implies that uptake cannot exceed demand, so that if demand is satisfied, excess nitrogen reaching the root surface in the transpiration stream is excluded. This phenomenon has been observed in experiments with plants grown in nutrient solutions, where total uptake levels off beyond a certain concentration of nitrogen in the solution even though transpiration continues (cf. Alberda, 1965) and also in the field where uptake by the vegetation levels off at high application rates (cf. Prins, Rauw & Postmus, 1981).

Distribution of nitrogen in the plant

The nitrogen taken up by the plant is partitioned between leaves, stems and roots in proportion to their relative demand. When supply cannot satisfy the total demand, roots do not have first priority despite their closeness to the source (van Keulen, 1981; van Dobben, 1963).

As the leaves age, some of their nitrogen can be transferred to tissues where an unsatisfied demand for nitrogen exists. If leaves die because nitrogen shortage has caused senescence, all the nitrogen except for an immobilisable residual amount in the senescing leaf, is translocated to the remaining live tissues. The residual level of nitrogen in the vegetative organs is a function of the development stage of the vegetation, young parts dying with a higher residual nitrogen concentration (Seligman, unpubl. data; Dilz, 1964). If leaves die for reasons other than nitrogen deficiency some of the translocatable nitrogen can be used to satisfy the nitrogen demand of other organs, primarily the stem. In this way the stem serves as a temporary store for nitrogen before translocation to the grain.

Translocation of nitrogen to the seed

Seeds receive most of their nitrogen in a reduced form, generally as amino acids, from the roots, leaves and stems (Donovan & Lee, 1978; Nair, Grover & Abrol, 1978). From various studies it appears that the rate of nitrogen accumulation in the grains can be considered constant during the linear phase of grain growth (Vos, 1981; Donovan & Lee, 1978; Sofield, Wardlaw, Evans & Zee, 1977). The rate of accumulation at any moment may be limited by the potential rate of accumulation in the grain (sink) (Donovan & Lee, 1977; 1978) or by the supply rate from the vegetative parts (source).

The rate of nitrogen depletion from the vegetative parts of the plants is fairly constant as long as the nitrogen concentration in the tissue is above a threshold level of around 0.01 kg kg^{-1} (Dalling, Boland & Wilson, 1976). Such a constant rate of depletion can be explained as withdrawal from a pool of amino-acids that is maintained at a more or less constant level, when calculated on an integrated daily basis (Hanson & Hitz, 1983). As the amino-acids are transferred from the vegetative tissue to the grain, storage or relatively stable proteins such as RuBPC–ase are mobilized, triggered by a rise in the level of proteolases at the onset of grain growth. The level of proteolases stays relatively high during grain filling and drops only as the grain approaches maturity (Dalling *et al.*, 1976).

As the concentration of nitrogen in the vegetative parts approaches the residual level the rate of depletion drops (Dalling *et al.*, 1976). The rate of transfer from the vegetative tissue and the uptake rate by the seeds are both dependent on temperature with a Q_{10} value of around 2 (Vos, 1981).

This process of nitrogen depletion in the vegetative parts and translocation to the grain is represented in the model by first defining the maximum nitrogen accumulation rate in the grains as a function of grain number and temperature. The potential export rate from the vegetative tissue is equal to the total nitrogen content above a residual level, multiplied by the relative turnover rate. The latter is influenced by temperature, the moisture status of the vegetation and the level of non-structural carbohydrate.

The actual rate of export is derived from the potential rate, taking into account the 'nitrogen activity' of the vegetative tissue, expressed as its average nitrogen concentration. This conceptualisation makes the notion of competition between vegetative tissue and seed explicit. Thus, if the vegetative tissue is 'active' and has a high nitrogen concentration, its competitive ability is high and the nitrogen is easily retained. The opposite is true in the reverse case. All of these relations are difficult to quantify on the basis of existing insights into the nitrogen economy of plants. Very little experimental work in this field appears to have been done, despite the importance of these processes for green area duration and hence grain yield on the one hand and grain nitrogen concentration, and hence baking quality, on the other. The temperature effect on nitrogen turnover in the plant is well-established and has already been discussed in treating maintenance respiration (Vos, 1981; Penning de Vries, 1975).

The effect of water stress on nitrogen turnover is difficult to disentangle from senescence. During water stressed conditions the nitrogen concentration of the leaf is usually higher (Halse, Greenwood, Lapins & Boundy 1969; Fischer & Kohn, 1966; Asana & Basu, 1963), perhaps because rapid senescence reduces the period available for translocation. Yet, as the tissues senesce, protein breakdown is accelerated. It can therefore be assumed that when the transpiration deficit increases nitrogen turnover increases.

Effect of nitrogen on organ formation

Leaf area formation. Nitrogen status mainly affects leaf area expansion through the amount of assimilate flowing to the leaves.

In a series of experiments on wheat, *Lolium rigidum* and *Lolium perenne* (Greenwood, 1966; Greenwood & Titmanis, 1966) the rate of leaf area expansion of the youngest expanding leaf was found to be linearly related to the total leaf nitrogen concentration at least for the 58 days duration of the experiments (Fig. 5.4).

However, whole canopy nitrogen concentration declines with plant development. Because the nitrogen distribution within the live leaf mass is not specified in the model, leaf expansion rate must be related not to the absolute nitrogen concentration in the leaf blades, but to the relative concentration on a scale running from the concentration found in severely depleted leaves up to a maximum concentration dependent on the development stage of the crop.

Under conditions of prolonged nitrogen stress, some of the nitrogen in older leaf blade and sheaths tissue is mobilised, translocated and resynthesised in new organs. The older tissues then die at a rate related to the average nitrogen concentration in the remaining leaf material. Stem tissue also dies under nitrogen shortage, particularly the leaf sheaths concurrent with the death of leaf tissue.

Effect of nitrogen on yield component formation. Formation of yield components, i.e. tillers, ears, spikelets, florets and grains is governed in the model by the assimilate supply necessary to create a viable organ. The effect of nitrogen status of the vegetation on organ formation is therefore mediated largely through its effect on gross assimilation. However, tiller formation is affected directly by the nitrogen status of the vegetation, as well as by assimilate availability (Yoshida & Hayakawa, 1970; Aspinall, 1961).

For defining the formation rate of the other organs the problem is that the effect of nitrogen status *per se* and the indirect effects *via* assimilate availability are difficult to disentangle and very little experimental work appears to have been done to separate the two effects. In one case where such an attempt was made, the direct effect of N status of the plant appeared to be small (Pinthus & Millet, 1978).

Fig. 5.4. The relation between the nitrogen concentration in the leaf and the relative rate of leaf area expansion (data points from three experiments).

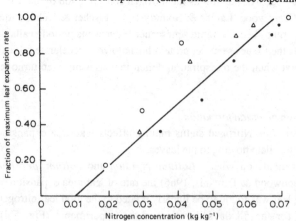

Some quantitive consequences of nitrogen metabolism

Yield-uptake and uptake-application relations

The processes and concepts outlined in the preceding section have been incorporated in a simulation model describing growth and yield of a spring wheat crop (van Keulen & Seligman, 1987). In this section some results from that model will be discussed, especially with respect to the consequences of the nitrogen economy of the vegetation. The model has been shown to be able to describe with reasonable accuracy the growth and yield of a spring wheat crop under semi-arid conditions in situations where either water or nitrogen is the limiting factor during part of the growing season.

The actual physical environment used in the simulation runs is not directly important, as the major concern is the sensitivity of the model to changes in crop characteristics related to the nitrogen economy. As detailed weather data were available for a site in the northern Negev desert of Israel, collected as part of a research project on actual and potential production in that region (van Keulen, Seligman & Benjamin, 1981), some of these data were used to drive the model.

In this region irrigated wheat cultivation is practised on a small scale. However, the effects of increased nutrient availability are much more obvious under conditions where other growth factors are close to the optimum. A set of simulation experiments was therefore executed assuming optimum moisture supply throughout the growth cycle of the crop.

First, a fertilizer experiment was simulated, assuming application of fertiliser at the time of sowing either in ammoniacal form or in the form of nitrate. The results presented in Fig. 5.5 show that the relationship between grain yield and nitrogen uptake was independent of the type of fertiliser applied. The maximum yield is approached asymptotically. At low uptake rates about 55 kg of grain is produced for each kg of nitrogen taken up. These results agree with experimentally determined response curves (van Keulen & van Heemst, 1982). The efficiency of nitrogen utilisation (E_N), expressed as kg grain produced per kg nitrogen taken up, is shown in Eqn 2.

$$E_N = 1/(0.01 + 0.004\ (W_s/W_g));\qquad\qquad(2)$$

W_s and W_g refer to the weight of straw and grain, respectively. The constants 0.01 and 0.004 are the minimum concentrations of nitrogen in grain and straw, respectively. The efficiency of nitrogen utilization increases with harvest index ($W_g/(W_s + W_g)$) and achieves a value of 55 at a harvest index of 0.33.

At increasing rates of uptake, the concentration of nitrogen in the harvested product increases and E_N declines. Finally, a plateau level is reached, where increased uptake does not lead to higher yields, because nitrogen is no longer the growth-determining factor, and yield is determined by the combined effect of radiation and temperature. At that point, there is also a tendency for decreasing harvest indices, because abundant nitrogen supply in the vegetative stage leads to luxurious vegetative growth and consequently to lower grain/straw ratios.

Nitrogen uptake was proportional to nitrogen application over the full range of application rates tested here, again in accordance with experimental results (van Keulen, 1986; van Keulen & van Heemst, 1982; van Keulen, 1977). However, a marked difference exists between the nitrate and the ammonium form of nitrogen. The recovery fraction, i.e. the ratio between nitrogen uptake and nitrogen application is 0.39 for the ammonium fertiliser and about 1.0 for the nitrate form.

The difference between the two forms of fertiliser is due to the semi-arid conditions for which the model was developed. Nitrate was assumed not to be subject to denitrification and leaching did not occur due to low rainfall. On the other hand ammonium was subject to volatilisation, which is especially important when fertiliser is applied early in the growing season before the rains have started.

Nitrogen turnover in the vegetative tissue

The proteins in the vegetative tissue of plants are not stable, but continuously degrade and have to be resynthesised. Unfortunately, only limited information is available on the rate of turnover of proteins, and most of that refers to very young leaves, so that it is questionable whether those data may be applied to mature or senescing leaves. In mature, functioning leaves, where the total protein content is more or less stable, the rate of protein turnover corresponds to breakdown and

Fig. 5.5. The simulated relation between total nitrogen uptake and grain yield, and that between nitrogen application and nitrogen uptake for a spring wheat crop grown in the northern Negev of Israel, for both ammonium (⊙, ⊗ and nitrate (●, ×) fertilisers (the cross and dot refer to two separate simulations).

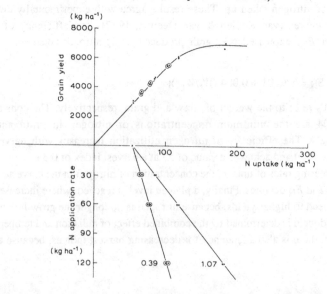

resynthesis of about 10% of the total protein each day (Penning de Vries, 1975; Huffaker & Peterson, 1974). In senescing leaves the rate may be higher, either because the 'activity' of the leaves decreases, or because the total protein content of the tissue is declining.

Translocation of nitrogen from the vegetative tissue to the developing grains can only take place if the nitrogenous components are in a form that can be transported, i.e. degraded proteins.

To test the sensitivity of crop performance to increased turnover rates, values of the relative turnover rate between 0.075 and 0.300 d^{-1} were tested in the model. The results, presented in Fig. 5.6, show that increasing turnover rates lead to both reduced grain and total dry matter yield, so that the harvest index remains more or less constant, except at the highest turnover rate where a slight drop in the harvest index occurs. These lower yields are the result of accelerated export of nitrogen from the vegetative tissue including the leaf blades, to the growing grain, leading to lower photosynthetic capacity and accelerated senescence.

The nitrogen harvest index, i.e. the ratio between total nitrogen in the grain at maturity and total nitrogen in the above ground material increased from 0.315–0.36 at the lowest turnover rate to 0.63–0.675 at the highest turnover rate, reflecting the more efficient translocation of nitrogen to the grains. The associated nitrogen concentration in the grains increases from 0.009–0.011 kg kg^{-1} to 0.024–0.029 kg kg^{-1}. These results thus indicate a negative relation between grain yield and nitrogen concentration in the grain (Fig. 5.7), in accordance with experimental results in which cultivars with high and low protein content in the grains have been compared (Kramer, 1979; Mesdag, 1979).

An interesting question arising from these results is whether genetic differences between high-protein and low-protein cultivars are the cause of lower yields, or are the result of differences in nitrogen turnover. In view of the interest in high nitrogen

Fig. 5.6. The simulated effect of the relative rate of nitrogen turnover in the vegetative tissue on *(a)* grain yield and *(b)* grain nitrogen concentration of a spring wheat crop grown in the northern Negev in two successive seasons.

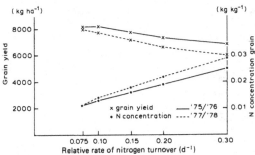

grains with an eye on baking quality an investigation into the genetic variability in nitrogen turnover rates in the crop canopy would be of interest.

The ontogenetic life span of the leaves

It appears that there is a strong interaction between the life span of the leaves and the crop's nitrogen economy. The ontogenetic life span of the leaves only plays a role under conditions of high nitrogen supply so that leaf death is not influenced by translocation of nitrogen from the vegetative tissues.

In the model leaves are assumed to function for 50 days at 15 °C, that is a thermal time of 750 °C d for a base temperature of 0 °C.

The two seasons illustrated in Fig. 5.6 differ in their reaction to increased leaf longevity: in 1976–77 longer leaf activity increased both total dry matter production and grain yield but hardly affected the harvest index. On the other hand in 1977–78 increased leaf longevity hardly affected either grain, or total dry matter production. Why there should be this difference between the two seasons is not clear.

An interesting observation from the model results is, however, that in both seasons the nitrogen harvest index increased substantially with increasing life span of the leaves. In 1976–77 from 0.50 at an average life span of 50 days to 0.60 at a life span of 70 days; for 1977–78 the values were 0.53 and 0.63, respectively. The longer active life of the leaves permits prolonged translocation of nitrogen to the growing grains and hence a more 'efficient use of the element. This is also borne out by the nitrogen concentration in the vegetative tissue at maturity, that was 0.014–0.015 kg kg^{-1} at a life span of 50 days and 0.011 when the life span of the leaves was 70 days.

Thus, greater leaf longevity creates a situation where, in contrast to the effect of increased nitrogen turnover, increased grain yields are accompanied by higher nitrogen concentrations in the grain. Whether such variability is present in the gene

Fig. 5.7. The simulated relationship between grain yield and nitrogen concentration in the grain of a spring wheat crop grown in the northern Negev for four successive seasons.

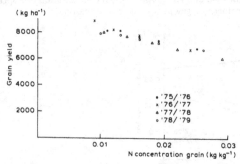

pool is difficult to judge, but the effect should be similar when growth under relatively cool conditions is compared with growth under relatively warm conditions (Spiertz & Ellen, 1978).

Concluding remarks

Despite the enormous amount of work that has been done on the subject and the associated proliferation of available literature, large gaps in our understanding of important basic processes still exist.

It would seem that only systems analysis, in which an attempt is made to 'put things together' within a coherent framework, is a research tool that leads to the explicit expression of such lack of knowledge.

Moreover, such an analysis may help in formulating goals for plant breeders: the data in Fig. 5.1 and 5.2 indicate that substantial variability exists in the relationship between assimilation and nitrogen concentration at the individual leaf level. That could be a trait to be pursued by plant breeders. Breeding for increased leaf longevity leading to higher grain and protein yields, could be another aim for plant breeders.

REFERENCES

Alberda, Th. (1965). The influence of temperature, light intensity and nitrate concentration on dry-matter production and chemical composition of *Lolium perenne* L. *Netherlands Journal of agricultural Science*, **13**, 335–60.

Angus, J.F. & Moncur, M.W. (1977). Water stress and phenology in wheat. *Australian Journal of agricultural Research*, **28**, 177–81

Asana, R.D. & Basu, R.N. (1963). Studies in physiological analysis of yield. VI. Analysis of the effect of water stress on grain development in wheat. *Indian Journal of Plant Physiology*, **6**, 1–13.

Aspinall, D. (1961). The control of tillering in the barley plant. I. The pattern of tillering and its relation to nutrient supply. *Australian Journal of biological Sciences*, **14**, 493–503.

Black, C.A. (1966). Crop yields in relation to water supply and soil fertility. In *Plant environment and Efficient Water Use*, eds. W.H. Pierre, D. Kirkham, J. Pesek & R. Shaw, pp. 177–206. American Society of Agronomy, Soil Science Society of America, Madison, Wisconsin

Boatwright, G.O. & Haas, H.J. (1961). Development and composition of spring wheat as influenced by nitrogen and phosphorus fertilisation. *Agronomy Journal*, **53**, 33–6.

Bolton, J.K. & Brown, R.H. (1980). Photosynthesis of grass species differing in carbon dioxide fixation pathways. V. Response of *Panicum maximum, Panicum milioides* and tall fescue (*Festuca arundinacea*) to nitrogen nutrition. *Plant Physiology*, **66**, 97–100.

Boon–Long, P., Egli, D.B. & Leggett, J.E. (1983). Leaf N and photosynthesis during reproductive growth in soybeans. *Crop Science*, **23**, 617–620.

Boote, K.J., Gallaher, R.N., Robertson, W.K., Hinson, K., & Hammond, L.C. (1978). Effect of foliar fertilization on photosynthesis, leaf nutrition, and yield of soybeans. *Agronomy Journal*, **70**, 787–91.

Brouwer, R. (1962). Nutritive influences on the distribution of dry matter in the plant. *Netherlands Journal of agricultural Science*. **10**, 399–408.

Brouwer, R. (1963). Some aspects of the equilibrium between overground and underground plant parts. *Jaarboek Instituut voor Biologisch en Scheikundig Onderzoek van Landbouwgewassen*, **1962**, 31–9.

Brouwer, R. (1965). Root growth of grasses and cereals. In: *The Growth of Cereals and Grasses.* eds. F.L. Milthorpe & J.D. Ivins. London: Butterworths. pp. 153–66.

Brouwer, R., Jenneskens, P.J. & Borggreve, G.J. (1962). Growth responses of shoots and roots to interruptions in the nitrogen supply. *Jaarboek Instituut voor Biologisch en Scheikundig Onderzoek van Landbouwgewassen*, **1961**, 29–36.

Brown, R.H. & Wilson, J.R. (1983). Nitrogen response of *Panicum* species differing in CO_2 fixation pathways. II. CO_2 exchange characteristics. *Crop Science*, **23**, 1154–9.

Burg, P.F.J. van, (1962). *Internal Nitrogen Balance, Production of Dry Matter and Ageing of Herbage and Grass.* Verslagen van landbouwkundige Onderzoekingen (Agricultural Research Reports) 68.12. Wageningen: Pudoc.

Campbell, C.A., Davidson, H.R. & McCaig, T.N., (1983). Disposition of nitrogen and soluble sugars in Manitou spring wheat as influenced by N fertilizer, temperature, and duration and stage of moisture stess. *Canadian Journal of Plant Science*, **63**, 73–90.

Campbell, C.A., Davidson, H.R. & Warder, F.G. (1977). Effects of fertilizer N and soil moisture on yield, yield components, protein content and N accumulation in the aboveground parts of sping wheat. *Canadian Journal of Soil Science*, **57**, 311–27.

Colman, R.L. & Lazenby, A. (1970). Factors affecting the response of some tropical and temperate grasses to fertilizer nitrogen. In *Proceedings 11th International Grassland Congress, Surfers Paradise*, ed. M.J.T. Norman, St Lucia: University of Queensland Press. pp. 392–7.

Cook, M.G. & Evans, L.T. (1983a). Nutrient responses of seedlings of wild and cultivated *Oryza* species. *Field Crops Research*, **6**, 205–18.

Cook, M.G. & Evans, L.T. (1983b). Some physiological aspects of the domestication and improvement of rice (*Oryza* spp.). *Field Crops Research*, **6**, 219–38.

Dalling, M.J., Boland, G. & Wilson, J.H. (1976). Relation between acid proteinase activity and redistribution of nitrogen during grain development in wheat. *Australian Journal of Plant Physiology*, **3**, 721–30.

Dantuma, G. (1973). Photosynthesis in leaves of wheat and barley. *Netherlands Journal of agricultural Science*, **21**, 188–98.

Dilz, K. (1964). *On the Optimum Nitrogen Nutrition of Cereals.* Verslagen van landbouwkundige Onderzoekingen (Agricultural Research Reports), 641. Wageningen: Pudoc.

Dobben, W.H. van (1960). Some observations on the nitrogen uptake of spring wheat and poppies in relation to growth. *Jaarboek Instituut voor Biologisch en Scheikundig Onderzoek van Landbouwgewassen*, **1959**, 93–105.

Dobben, W.H. van (1962a). Influence of temperature and light conditions on dry-matter distribution, development rate and yield in arable crops. *Netherlands Journal of agricultural Science*, **10**, 377–89.

Dobben, W.H. van (1962b). Nitrogen uptake of spring wheat and poppies in relation to growth and development. *Jaarboek Instituut voor Biologisch en Scheikundig Onderzoek van Landbouwgewassen*, **1961**, 45–60.

Dobben, W.H. van (1963). The distribution of dry matter in cereals in relation to nitrogen nutrition. *Jaarboek Instituut voor Biologisch en Scheikundig Onderzoek van Landbouwgewassen*, **1962**, 77–89.

Donovan, G.R. & Lee, J.W. (1977). The growth of detached wheat heads in liquid culture. *Plant Science Letters*, **9**, 107–13.

Donovan, G.R. & Lee, J.W. (1978). Effect of nitrogen source on grain development in detached wheat heads in liquid culture. *Australian Journal of Plant Physiology*, **5**, 81–7.

Fischer, R.A. & Kohn, G.D. (1966). The relationship of grain yield to vegetative growth and post-flowering leaf area in the wheat crop under conditions of limited soil moisture. *Australian Journal of agricultural Research*, **17**, 281–95.

Gajri, P.R. & Prihar, S.S. (1985). Rooting, water use and yield relations in wheat on loamy sand and sandy loam soils. *Field Crops Research*, **12**, 115–32.

Goudriaan, J. & van Keulen, H. (1979). The direct and indirect effects of nitrogen shortage on photosynthesis and transpiration in maize and sunflower. *Netherlands Journal of agricultural Science*, **27**, 227–34.

Goudriaan, J. & H.H. van Laar, H.H. (1978). Relations between leaf resistance, CO_2-concentraion and CO_2-assimilation in maize, beans, lalang grass and sunflower. *Photosynthetica*, **12**, 241–9.

Greenwood, D.J. (1982). Modelling of crop response to nitrogen fertilizer. *Philosophical Transactions of the Royal Society, Series B* **296**, 351–62.

Greenwood, E.A.N. (1966). Nitrogen stress in wheat – its measurement and relation to leaf nitrogen. *Plant and Soil*, **24**, 279–88.

Greenwood, E.A.N. & Titmanis, Z.V. (1966). The effect of age on nitrogen stress and its relation to leaf nitrogen and leaf elongation in a grass. *Plant and Soil*, **24**, 379–89.

Halse, N.J., Greenwood, E.A.N., Lapins, P. & Boundy, C.A.P. (1969). An analysis of the effects of nitrogen deficiency on the growth and yield of a western Australian wheat crop. *Australian Journal of agricultural Research*, **20**, 987–98.

Hanson, A.D. & Hitz, W.D. (1983). Whole-plant response to water deficits: Water deficits and the nitrogen economy. In: *Limitations of Efficient Water Use in Crop Production*. eds. H.M. Taylor, W.R. Jordan & T.R. Sinclair. American Society of Agronomy Monograph, Madison, Wisconsin: American Society of Agronomy. pp. 331–43.

Hochman, Z. (1982). Effect of water stress with phasic development on yield of wheat grown in a semi-arid environment. *Field Crops Research*, **5**, 55–67.

Huffaker, R.C. & Peterson, L.W. (1974). Protein turnover in plants and possible means of its regulation. *Annual Review of Plant Physiology*, **25**, 363–92.

Ishihara, K., Ebara, H., Hirawasa, T. & Ogura, T. (1978). The relationship between environmental factors and behaviour of stomata in the rice plants. VII. The relation between nitrogen content in leaf blades and stomatal aperture. *Japanese Journal of Crop Science*, **47**, 664–73.

Keulen, H. van (1975). *Simulation of Water Use and Herbage Growth in Arid Regions*. Simulation Monographs. Wageningen, Pudoc.

Keulen, H. van (1977). *Nitrogen Requirements of Rice with Special Reference to Java*. Contributions of the Central Research Institute for Agriculture Bogor 30. Bogor, Indonesia: Central Research Institute for Agriculture.

Keulen, H. van (1981) Modelling the interaction of water and nitrogen. *Plant and Soil*, **58**, 205–29.

Keulen, H. van (1986). Crop yield and nutrient requirements. In *Modelling of Agricultural Production: Weather, Soils and Crops*. eds. H. van Keulen & J. Wolf. Simulation Monographs. Wageningen : Pudoc. pp 155–181.

Keulen, H. van & Seligman, N.G. (1987). *Simulation of Water Use, Nitrogen Nutrition and Growth of a Spring Wheat Crop*. Simulation Monographs. Wageningen:Pudoc.

Keulen, H. van & van Heemst, H.D.J.(1982) *Crop Response to the Supply of Macronutrients*. Verslagen van landbouwkundige Onderzoekingen (Agricultural Research Reports) 916. Wageningen:Pudoc.

Keulen, H. van, Seligman, N.G. & Goudriaan, J. (1975). Availability of anions in the growth medium to roots of an actively growing plant. *Netherlands Journal of agricultural Science*, **23**, 131–8.

Keulen, H. van, Seligman, N.G. & Benjamin, R.W. (1981). Simulation of water use and herbage growth in arid regions – A re-evaluation and further development of the model 'Arid Crop'. *Agricultural Systems*, **6**, 159–93.

Khan, M.A. & Tsunoda, S. (1970a). Evolutionary trends in leaf photosynthesis and related leaf characters among cultivated wheat species and its wild relatives. *Japanese Journal of Breeding*, **20**, 133–40.

Khan, M.A. & Tsunoda, S. (1970b). Differences in leaf photosynthesis and leaf transpiration rate among six commercial wheat varieties of west Pakistan. *Japanese Journal of Breeding*, **20**, 344–50.

Kramer, Th. (1979). Yield-protein relationship in cereal varieties. In: *Crop Physiology and Cereal Breeding*, eds. J.H.J. Spiertz & Th. Kramer, Wageningen:Pudoc. pp. 161–5.

Lof, H. (1976). *Water Use Efficiency and Competition Between Arid Zone Annuals, Especially the Grasses* Phalaris minor and Hordeum murinum. Verslagen van landbouwkundige Onderzoekingen (Agricultural Research Reports) 853. Wageningen:Pudoc.

Loustalot, A.J., Gilbert, S.G. & Drosdoff, A. (1950). The effect of nitrogen and potassium levels in tung seedlings on growth, apparent photosynthesis and carbohydrate composition. *Plant Physiology*, **25**, 394–412.

Lugg, D.G. & Sinclair, T.R. (1981). Seasonal changes in photosynthesis of field-grown soybean leaflets. 2. Relation to nitrogen content. *Photosynthetica*, **15**, 138–44.

Lupton, F.G.H., Oliver, R.H., Ellis, F.B., Barnes, B.T., Howse, K.R., Welbank, P.J., Taylor, P.J. (1974). Root and shoot growth of semi-dwarf and taller winter wheats. *Annals of applied Biology*, **72**, 129–44.

Marshall, B. (1978). *Leaf and Ear Photosynthesis of Winter Wheat crops*. PhD Thesis, University of Nottingham.

Martin, R.J. & Dougherty, C.T. (1975). Diurnal variation of water potential of wheat under contrasting weather conditions. *New Zealand Journal of agricultural Research*, **18**, 145–8.

McLean, E.O. (1957). Plant growth and uptake of nutrients as influenced by levels of nitrogen. *Soil Science Society of America Proceedings*, **21**, 219–22.

McNeal, F.H., Berg, M.A., Watson, C.A. (1966). Nitrogen and dry matter in five spring wheat varieties at successive stages of development. *Agronomy Journal*, **58**, 605–8.

Mesdag, J. (1979). Genetic variation in grain yield and protein content of spring wheat (*Triticum aestivum* L.). In *Crop Physiology and Cereal Breeding*, eds. J.H.J. Spiertz & Th. Kramer, Wageningen:Pudoc., P.J. pp. 166–7.

Mooney, H.A., Ferrar, P.J. & Slatyer, R.O. (1978). Photosynthetic capacity and carbon allocation patterns in diverse growth forms of Eucalyptus. *Oecologia (Berlin)*, **36**, 103–11.

Nair, T.V.R., Grover H.L. & Abrol, Y.P. (1978). Nitrogen metabolism of the upper three leaf blades of wheat at different soil nitrogen levels. II. Protease activity and mobilization of reduced nitrogen to the developing grains. *Physiologia Plantarum*, **42**, 293–300.

Nevins, D.J. & Loomis, R.S. (1970). Nitrogen nutrition and photosynthesis in sugar beet. *Crop Science*, **10**, 21–5.

Os, A.J. van (1967). The influence of nitrogen supply on the distribution of dry matter in spring rye. *Jaarboek Instituut voor Biologisch en Scheikundig Onderzoek van Landbouwgewassen*, **1966**, 51–65.

Osman, A.M. & Milthorpe, F.L. (1971). Photosynthesis of wheat leaves in relation to age, illuminance and nutrient supply. II. Results. *Photosynthetica*, **5**, 61–70.

Osman, A.M., Goodman, P.J. & Cooper, J.P. (1977). The effects of nitrogen, phosphorus and potassium on rates of growth and photosynthesis in wheat. *Photosynthetica*, **11**, 66–75.

Penning de Vries, F.W.T. (1974). Substrate utilization and respiration in relation to growth and maintenance in higher plants. *Netherlands Journal of agricultural Science*, **22**, 40–4.

Penning de Vries, F.W.T. (1975). The cost of maintenance processes in plant cells. *Annals of Botany (London)*, **39**, 77–92.

Penning de Vries, F.W.T., Brunsting, A.H.M & van Laar, H.H. (1974). Products, requirements and efficiency of biosynthesis: a quantitative approach. *Journal of theoretical Biology*, **45**, 339–77.

Pinthus, M.J. & Millet, E. (1978). Interactions among number of spikelets, number of grains and grain weight in the spikes of wheat (*Triticum aestivum* L.). *Annals of Botany* **42**, 839–48.

Prins, W.H., Rauw, G.J.G., Postmus, J. (1981). Very high application of nitrogen fertilizer on grassland and residual effects in the following season. *Fertilizer Research* **2**, 309–27.

Radin, J.W. (1981). Water relations of cotton plants under nitrogen deficiency. IV. Leaf senescence during drought and its relation to stomatal closure. *Physiologia Plantarum*, **51**, 145–9.

Radin, J.W. (1983). Control of plant growth by nitrogen: difference between cereals and broadleaf species. *Plant Cell and Environment* **6**, 65–8.

Radin, J.W. & Ackerson, R.C. (1981). Water relations of cotton plants under nitrogen deficiency. III. Stomatal conductance, photosynthesis, and abscisic acid accumulation during drought. *Plant Physiology*, **67**, 115–19.

Radin, J.W. & Boyer, J.S. (1982). Control of leaf expansion by nitrogen nutrition in sunflower plants: role of hydraulic conductivity and turgor. *Plant Physiology* **69**, 771–5.

Radin, J.W. & Parker, L.L.(1979a). Water relations of cotton plants under nitrogen deficiency. I. Dependence upon leaf structure. *Plant Physiology*, **64**, 495–8.

Radin, J.W. & Parker, L.L. (1979b). Water relations of cotton plants under nitrogen stress. II. Environmental interactions on stomata. *Plant Physiology*, **64**, 499–501.

Ryle, G.J.A. & Hesketh, J.D. (1969). Carbon dioxide uptake in nitrogen-deficient plants. *Crop Science*, **9**, 451–4.

Seligman, N.G., Loomis, R.S., Burke, J. & Abshahi, A. (1983). Nitrogen nutrition and phenological development in field-grown wheat. *Journal of agricultural Science*, **101**, 691–7.

Seligman, N.G., van Keulen, H., Yulzari, A., Yonathan, R., Benjamin, R.W. (1976). *The Effect of Abundant Nitrogen Fertilizer Application on the Seasonal Change in Mineral Concentration in Annual Mediterranean Pasture Species.* Preliminary Report 754.Bet Dagan, Israel: Division of Scientific Publications.

Shimshi, D.(1970a). The effect of nitrogen supply on some indices of plant–water relations of beans (*Phaseolus vulgaris* L.). *New Phytologist*, **69**, 413–24.

Shimshi, D. (1970b). The effect of nitrogen supply on transpiration and stomatal behaviour of beans (*Phaseolus vulgaris* L.). *New Phytologist*, **69**, 405–13.

Shimshi, D. & Kafkafi, U. (1978). The effect of supplemental irrigation and nitrogen fertilisation on wheat (*Triticum aestivum* L.). *Irrigation Science*, **1**, 27–38.

Sofield, I., Wardlaw, I.F., Evans, L.T., Zee, S.Y. (1977). Nitrogen, phosphorus and water contents during grain development and maturation in wheat. *Australian Journal of Plant Physiology*, **4**, 799–810.

Spiertz, J.H.H. & Ellen, J.(1978). Effects of nitrogen on crop development and grain growth of winter wheat in relation to assimilation and utilisation of assimilates and nutrients. *Netherlands Journal of agricultural Science*, 26, 210–31.

Sunderland, N. (1960). Cell division and expansion in the growth of the leaf. *Journal of experimental Botany* 11, 68–80.

Takeda, T. (1961). Studies on the photosynthesis and production of dry matter in the community of rice plants. *Japanese Journal of Botany*, 17, 403–37.

Tanner, C.B. & Sinclair, T.R. (1983). Efficient water use in crop production: Research or re-search. In *Limitations of Efficient Water Use in Crop Production*. eds. H.M. Taylor, W.R. Jordan & T.R. Sinclair. American Society of Agronomy Monograph. Madison, Wisconsin: American Society of Agronomy, pp. 1–27.

Viets, F.G. Jr. (1962). Fertilizers and the efficient use of water. *Advances in Agronomy*, 14, 223–64.

Vos, J. (1981). *Effects of Temperature and Nitrogen Supply on Post-floral Growth of Wheat; Measurements and Simulations*. Verslagen van Landbouwkundige Onderzoekingen (Agricultural Research Reports) 911. Wageningen:Pudoc.

Wilson, J.R. (1975a). Influence of temperature and nitrogen on growth, photosynthesis and accumulation of non-structural carbohydrate in a tropical grass, *Panicum maximum var. trichoglume*. *Netherlands Journal of agricultural Science*, 23, 48–61.

Wilson, J.R. (1975b). Comparative response to nitrogen deficiency of a tropical and temperate grass in the interrelation between photosynthesis, growth, and the accumulation of non-structural carbohydrate. *Netherlands Journal of agricultural Science*, 23, 104–12.

Wilson, J.R. & Haydock, (1971). The comparative response of tropical and temperate grasses to varying levels of nitrogen and phosphorus nutrition. *Australian Journal of agricultural Research*, 22, 573–87.

Wit, C.T. de (1958). *Transpiration and Crop Yields*. Verslagen van landbouwkundige Onderzoekingen (Agricultural Research Reports) 64.8 Wageningen:Pudoc.

Wit, C.T. de et al.(1978). *Simulation of Assimilation, Respiration and Transpiration of Crops*. Simulation Monographs. Wageningen:Pudoc.

Woledge, J. & Pearse, P.J. (1985). The effect of nitrogenous fertilizer on the photosynthesis of leaves of a ryegrass sward. *Grass and Forage Science* 40, 306–9.

Wong, S.C. (1979). Elevated atmospheric partial pressure of CO_2 and plant growth.I. Interactions of nitrogen nutrition and photosynthetic capacity in C_3 and C_4 plants. *Oecologia (Berlin)*, 44, 68–74.

Wong, S.C., Cowan, I.R. & Farquhar, G.D. (1979). Stomatal conductance correlates with photosynthetic capacity. *Nature*, 282, 424–6.

Yoshida, S. & Coronel, V.(1976). Nitrogen nutrition, leaf resistance and leaf photosynthetic rate of the rice plant. *Soil Science and Plant Nutrition*, 22, 207–11.

Yoshida, S. & Hayakawa,Y. (1970). Effects of mineral nutrition on tillering of rice. *Soil Science and Plant Nutrition*, 16, 186–91.

JOHN L. HARPER

6. Canopies as populations

In most studies of crop canopies or of the foliage of single plants, all leaves are treated as if they have the same properties. This is done so that we may make generalisations about the ways in which plant or crop growth rates may be interpreted as a function of leaf area. There is no gainsaying that this approach with its underlying assumption has been profitable. Concepts such as leaf area index (LAI) and net assimilation rate (NAR) have contributed greatly to our understanding of how a photosynthetic surface contributes to determining the growth rate of plants. However, the assumption is false.

The leaves on a plant or in a crop form a population, an assemblage of things that can be counted, and they are manifestly not all the same. Their heterogeneity derives in part from the fact that they (like a population of rabbits in a field or of blue tits in a woodland) are not of the same age and change their properties as they age. They are also borne in different positions relative to each other and their positions determine which leaves shade which. The positions that they occupy in a canopy are also related to their age – in general, young leaves are found in the fringes of a canopy with older ones in their shade.

Population biologists have much experience of studying the behaviour of age-structured populations and the aim of this chapter is to explore how far the study of populations (demography) may contribute to the study of plant canopies.

The growth of populations

The fundamental equation of population biology relates the numbers in a population at time $t+1$ to the numbers present at time t. The numbers of, say, rabbits in a field at time $t+1$ is the number present at time t plus births in the time interval, minus deaths, plus the number of immigrants to the field, minus the emigrants from it.

$$N^{t+1} = N^t + B - D + I - E. \tag{1}$$

For a population of leaves on a fixed rooted plant or within a quadrat of a crop in the field, immigration and emigration are usually impossible or can be ignored. They need to be considered only in quadrat studies when a plant may produce leaves that spread beyond the quadrat (emigrate) or a plant outside the quadrat may extend into it (immigrate). The risk of assuming that emigration and immigration are equal is much less for a population of leaves than for a population of rabbits! The risk is probably greatest if the plant is rhizomatous or stoloniferous and it would be unwise to ignore immigration and emigration in a quadrat study of the demography of a clonally spreading weed within a crop. In general, however, we can usually ignore I and E. The growth of a canopy can then be described by the simpler equation:

$$N^{t+1} = N^t + B - D \ . \tag{2}$$

Birth

The use of the term 'birth' for a leaf is not as facile as appears at first sight. As in a mammal, birth is a stage delayed after conception when the largely preformed embryo emerges from the mother. A student of canopy demography can usually ignore the phases between the initiation of a leaf primordium and its emergence from the parent bud. It should not be forgotten, however, that much abortion may occur in populations of leaves before the phase of emergence, i.e. birth. Most buds die before they expand any leaves and so most leaves die before they are 'born'.

The time at which an expanding young leaf may be said to be 'born' is of course arbitrary. In studies of *Linum usitatissimum* which are described in this paper, leaves produced at different times were identified by slipping segments of coloured drinking straw over the stem apex. A leaf that had expanded sufficiently to stop the straw from slipping further down the stem was considered 'born'. In grasses, the emergence of a leaf tip from the enclosing sheath may be considered to be the moment of birth. These measures are no more arbitrary than the decision to define the birth of a kangaroo as the time the embryo first enters the mother's pouch or the time it first leaves it.

The birth rate of leaves in cereal crops appears to be essentially a simple function of thermal time with the time interval between the birth of successive leaves (the phyllochron) being constant in thermal time units. Gallagher (1979) reported values of about 110 $^\circ$C (base temperature = 0 $^\circ$C) for leaves of winter wheat (*Triticum aestivum*) and spring barley (*Hordeum vulgare*). Gallagher's generalisation has been generally confirmed by others, but the 'best' base temperature for the calculation may change with time (4–5 $^\circ$C for September sowings and less than 0 $^\circ$C for spring sowings (Kirby, Appleyard & Fellowes, 1982). There is also evidence from winter wheat and from winter and spring barley that the rate of change of daylength at the time of the crop's emergence affects the phyllochron in such a way that the canopy of late sown crops tends to catch up with those sown earlier (Baker, Gallagher & Monteith, 1980; Ellis & Russell, 1984).

Sackville–Hamilton (pers. comm.) followed the birth of leaves on individual stolons of white clover (*Trifolium repens*) in an old permanent pasture in N. Wales throughout the year. He used maximum and minimum air temperature, soil temperature, dry and wet bulb temperatures, rainfall, hours of sunlight and daylength, as cumulated values, squares, cubes, cross-products and differences between plants (presumed to reflect genotypic or local micro-environmental effects) as predictors of leaf birth rate in a multiple regression model. Of these variables, accumulated soil temperature accounted for 79% of the variation in leaf birth rate. The annual birth rate of leaves per shoot increased in this simple model by 4.75 leaves per year for each 1 °C rise in temperature. The statistical procedure allowed him to derive the base temperature of 3.5 °C from the data and not simply to assume it. The residual variance was reduced by 30% after including further climatic terms and derivatives. When calculated differences between the behaviour of different stolons were included in the model it left only 3% of the variation in leaf birth rate unaccounted for.

Other species may have leaf birth rates that are determined in more complex ways. Two broad categories can be recognised among temperate trees and perennial herbs. In plants with determinate growth, leaves preformed in the bud in the previous year are born in a rapid spring flush and no more leaves are born in that season, for example pines (*Pinus spp.*) and daffodils (*Narcissus spp.*). In others, for example birch (*Betula sp.*) and dandelions (*Taraxacum spp.*), an initial flush of spring births of leaves preformed in the winter buds is followed by a longer drawn out sequence of initiations and births in the same season. In many tropical trees, for example *Theobroma cacao*, seasonal or apparently erratic bursts of leaf births occur throughout the year.

Once floral initiation has begun on a cereal shoot, no more leaf primordia are formed and the last leaf is born some time later. In clover, floral primordia develop from axillary buds rather than apical meristems, so that the rate at which leaves are born appears to be a function of climate only (mainly temperature) throughout the year.

Leaf birth can be expressed in a variety of ways. In the clover and cereal examples, birth rate was considered as leaves born per shoot per unit of time. This takes no account of the development of branches and the birth of leaves on them. In these species whether an axillary bud develops and contributes its own leaf birth rate to the plant is clearly not tightly programmed genetically. The number of tillers per plant of a cereal or branches per plant of a clover stolon is highly responsive to local environmental conditions, especially light and nutrient resources. If branches (tillers) do develop on the cereal, their development is closely linked to the birth of leaves on the main shoot. In winter wheat, Tiller 1 (T1) starts to develop at the same time as leaf 4 on the main shoot and tiller T2 at the same time as leaf 5 (Masle–Meynard & Sebillotte, 1981) and the phyllochrons are the same for the main shoot and its tillers (Klepper, Rickman & Peterson, 1982). So long as axillary buds continue to develop into branches, each bearing its own leaves, the leaf population of the plant will tend to exponential growth, though each individual shoot continues to give birth to new leaves in a roughly linear sequence over thermal time. Thus we obtain quite a different picture of leaf birth rates per shoot apex and per plant.

It is sometimes convenient to express leaf birth rates per area of land when it is relevant to interpreting changes in leaf area index. Leaf birth can also be expressed as a fraction of the population of leaves on a plant or on an area of land.

From the moment of birth a leaf is committed to the remorseless forces of mortality – it is on its way to death. The analysis of death rates and the analysis of their causes are as important a part of demographic analysis as is the study of births.

Death

The demography of leaf death is quite easy to describe and analyse except that (as with birth) there is an arbitrariness involved in determining precisely when it occurs. In plant science, as in medicine, there is room for much argument about just when the subject is considered to be dead. In practice, it makes little difference which convention is adopted, provided it is adhered to throughout a study. It is often easiest to regard a leaf as dead when more than half of its area is yellowed or browned, withered or dried.

If the dates of birth of leaves are known it becomes possible to describe the age structure of the leaf population on a plant or on an area of ground. The only way we can obtain direct measurements of this sort is by marking and following the individual leaves within a canopy from birth to death. In practice it is almost always convenient to group leaves into 'cohorts' born within specific time intervals and to mark them with different coloured tags or paint spots or with some other device. It is then possible to determine the death rate of leaves that were born at different times. Death rates are conventionally expressed as the decline in numbers over time from an ideal starting population of 1000 leaves. The data are expressed on semi-logarithmic graphs so that rates of change can be compared.

Demographers recognise three model survivorship curves which represent variations along a continuum (Fig. 6.1). The survivorship curves for leaves generally conform to Type 1 – i.e. they have a juvenile period during which they have only a slight risk of dying but the risk of death increases rapidly with age.

An example of leaf survivorship curves comes from a study of the behaviour of a

Fig. 6.1. Three model survivorship curves.

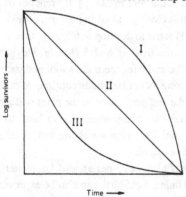

sward of *Lolium perenne* (Peters, 1980). The sward was sown in January 1977 in an unheated glasshouse and, in the treatment described here, the growing sward was cut to a height of 10 cm above soil level at monthly intervals. In this study a leaf was considered dead when all its tissue had become brown–yellow.

The main finding appears surprising at first sight (Fig. 6.2). Leaves born in the winter have long life expectancy and this shortens in the spring and is shortest for leaves born at the beginning of summer. Of the leaves born in November, half had died after 8 weeks and in contrast, of leaves born in May half had died after only 4 weeks. The winter born leaves spent the first 6–7 weeks of their lives with scarcely any risk of death, whereas among those born in May, deaths were occurring even in the first week of their lives.

There is a crude relationship between the birth rate of new leaves and the death rate of those already present in the sward. Leaves are born only slowly in the winter, but those that are produced live long. Leaves are produced more rapidly in the summer (especially in May) but have but a short time to live. The relationship between birth and death rates is shown in Fig. 6.3 for another of Peters' studies in which the demography of leaves was studied in an old permanent pasture. The correlation is particularly strong for white clover where for every new leaf added at the end of a stolon an older one died – usually in the same week. It is difficult to avoid the impression that the birth of new leaves 'causes' the death of old or *vice versa*. In these studies the birth and death rates of leaves were determined on individual grass tillers or on the main shoots of clover stolons – they do not take into account births and deaths on new tillers or stolon branches. If we consider the demography of the canopy of clover or grass leaves per unit area of an established sward, the birth and death rates on individual tillers or stolon axes are such as to maintain the *status quo*. Any significant increases in the number of leaves in the canopy must come from new branches, not from more leaves born per shoot.

Fig. 6.2. Survivorship curves for cohorts of leaves of *Lolium perenne* born in different months in a sward maintained under glasshouse conditions (from Peters, 1980).

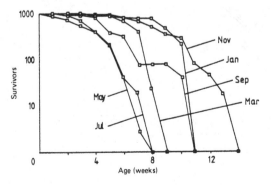

Fig. 6.3. The relationship between relative birth rates and relative death rates of leaves in a permanent pasture. The relative rates are expressed as the number of leaves born (or dying) as a proportion of the total number present on a shoot or stolon: (a) *Lolium perenne*; (b) *Trifolium repens*.

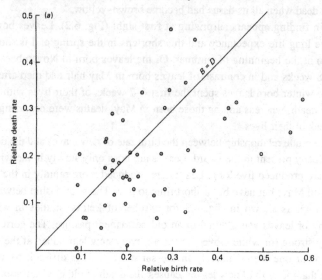

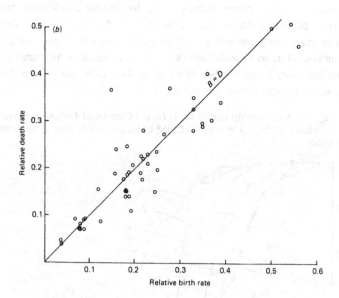

Statistical procedures for fitting curves and comparing survivorship have been described by Pyke & Thompson (1986). Peters used the Weibull function which was developed for 'time to failure' studies in engineering (e.g. metal fatigue) and as a model of human survivorship (Gehan & Siddiqui, 1973), seed vigour (Bonner & Dell, 1976) and the survivorship of a range of organisms (Pinder, Wiener & Smith, 1978). Its strong descriptive power for leaf survivorship is shown by the r^2 values which usually exceed 0.90 and never fell below 0.82.

The survivorship curves for leaves of *Lolium perenne* (Fig. 6.2) are very similar to those obtained by Williamson (1976) for five species of grass in an ungrazed chalk grassland and of Clark (1980) for *Catapodium rigidum* and *C. murinum* grown in pots under controlled conditions.

A canopy of leaves has an age structure defined by the proportions of leaves of different ages within it. The way in which the age structure of a grass sward may change during the year is shown in Fig. 6.4 for the glasshouse population of *Lolium perenne* studied by Peters. The canopy was composed mainly of young leaves during May, June and July when birth and death rates were high but in the overwintering population nearly 20% of the leaves were more than 9 weeks old.

Factors affecting leaf demography
Population density and shading

Linum usitatissimum is a convenient plant for the study of leaf demography. The plant's life cycle has almost no effect on the size of individual leaves and new leaves emerge rapidly from apical buds, 3 apex^{-1} d^{-1} at the peak of the growing season. It is easy to mark successive cohorts of leaves by slipping colour-coded plastic rings over the shoot apex so that they come to rest on the first expanded leaf (and so define leaves that have been 'born'). Isolated plants branch from the cotyledonary axils and may also branch from the axils of the first true leaves. If

Fig. 6.4. The changing age structure of the leaf population in a sward of *Lolium perenne* maintained under glasshouse conditions and cut to 10 cm height at 4-week intervals (from Peters, 1980).

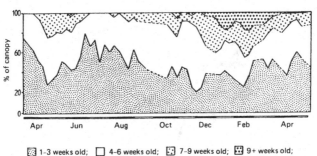

■ 1–3 weeks old; □ 4–6 weeks old; ▨ 7–9 weeks old; ▨ 9+ weeks old;

resources are depleted, for example when plants are grown close together, fewer branches are produced and those that develop remain small.

Bazzaz & Harper (1977) grew *L. usitatissimum* in John Innes Compost No 1 in trays (8×16×20cm) at densities of 3, 9 and 30 per tray. The experiment was performed in a glasshouse set at approximately 20 °C during the day and 15 °C at night. The plants in half of the trays were shaded to 50% of full daylight with polythene netting. Plastic rings were slipped over the stem apices at 3–4-day intervals and the leaves between two rings were treated as an even-aged cohort. A leaf was considered dead when it had completely yellowed. The effects of treatments on the survivorship of the leaves are shown in Fig. 6.5 and the age structure of the populations is shown for three dates in Fig. 6.6.

The effect of density was to reduce branching and thus to reduce the number of apices contributing to the birth rate of the leaf population. This is shown in Fig. 6.7 in which the number of leaves per cohort rose steeply in the uncrowded plants as additional growing points were added to the population. (Note that in this case leaf birth and death are shown *per plant*, in contrast to Figs. 6.2 and 6.3 in which they were expressed per shoot.) The leaves on crowded plants were not only produced more slowly, they also died more quickly. The effect of shading was to prolong the lives of the leaves, at all densities, by 15–20 days.

In these studies the onset of leaf death was *not* predicted by the onset of flowering. In full light leaf death started about 15 days *after* the first flower opened at the lowest plant density and about 5 days *before* flowers opened at the highest density. In all cases leaf death occurred as a synchronous wave passing up each plant and, regardless of their age, it was the lower leaves that died first (leaves produced on side branches were younger than leaves on the main axis at the same height, but started to die at the same time).

Fig. 6.5. The survivorship curves for leaves of *Linum usitatissimum* grown at three densities and in two shading regimes. *A*, *B* and *C* are increasing densities at full light intensity and *D*, *E* and *F* are increasing densities under 50% shading (from Bazzazz Harper, 1977).

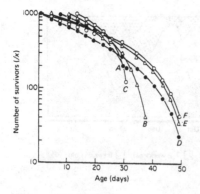

The strongest predictor of the onset of leaf death was the biomass per tray. Leaves started to die when the total dry weight per plant was about 13 g, regardless of LAI, the leaf area per plant, the population density or the light intensity. The plants behaved as if some resource present in the soil of a tray limited the 'carrying capacity' of the canopy population. Leaf death started when this carrying capacity was reached.

Mineral nutrition

One result of the studies of the effects of density and shading on the leaf demography of *Linum* was to focus attention on the effects of mineral nutrition on canopy dynamics. *Linum* was again chosen as the test species and seeds were sown in acid washed sand in 100 mm plastic pots. The seeds were sown on 23 June 1978 and on 3 July the seedlings were thinned to one per pot. The pots were covered with black plastic with a central hole to accommodate the seeding – the plastic was to prevent

Fig. 6.6. The changing age structure of populations of leaves of *Linum usitatissimum* grown in full sunlight and under 50% shading (from Bazzaz & Harper, 1977).

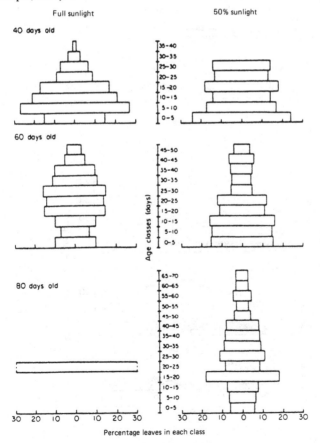

Percentage leaves in each class

Fig. 6.7. The development of the canopy in populations of *Linum usitatissimum* grown at densities of 3 (O), 9(Δ) and 30 (□) plants per tray in full daylight (*a*), and under 50% shading (*b*). The birth, life span and death of successive leaf cohorts are shown. The letters *c*, *f* and *d* refer respectively to the time of cotyledon death, the initiation of flowering and the onset of leaf death (from Bazzaz & Harper, 1977).

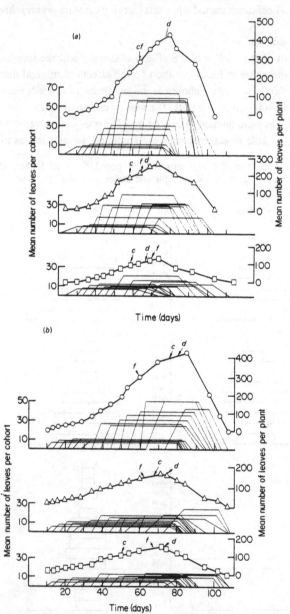

algal growth and exclude contamination by dust. Nutrients were provided initially to the bottom of the pots with a tube and reservoir system but as the plants grew larger the nutrient solution was applied directly to the surface of the sand. The nutrient solutions supplied were Long Ashton nutrient solution (Hewitt, 1966) complete or with necessary omissions or substitutions. The experiment included a distilled water control in which all the growth made by the plants depended on seed reserves. Cohorts of leaves were marked as explained earlier.

The results of this experiment are described only briefly here to show how the techniques of leaf demography may be used to analyse the effects of nutrient deficiency and to quantify features that are usually described only qualitatively or as colour photographs (e.g. Wallace, 1944).

Fig. 6.8 shows birth and death in successive cohorts of leaves of plants receiving complete Long Ashton solution and plants deprived of potassium or of calcium. The dynamics of the leaf populations are clearly quite different. The potassium deficient plants (Fig. 6.8(*b*)) continued to give birth to new leaves for as long as the plants that received full nutrients (100 days, see Fig. 6.8(*a*)) but the leaves on potassium deficient plants were very short lived. Thus the plants receiving full nutrient supply accumulated a persistent canopy of long-lived leaves, whereas the potassium deficient

Fig. 6.8. The development of the leaf population on plants of *Linum usitatissimum* grown in sand culture receiving (*a*) full Long Ashton nutrient solution, (*b*) nutrient solution omitting potassium, and (*c*) nutrient solution omitting calcium. The diagram shows the birth, life span and death of successive leaf cohorts.

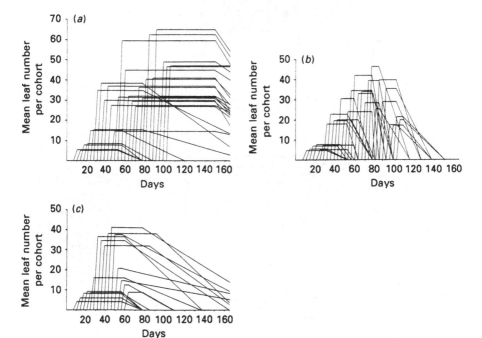

plants carried a canopy of ephemeral leaves made up of a tuft of young leaves at the top of the shoot and an extended zone of death below.

Fig. 6.8(c) shows a comparable diagram for the leaf dynamics of calcium deficient plants. In this case the birth of new leaves ceased after 60 days, and although the leaves in each cohort started to die soon after they were formed, they did so much more slowly than in potassium deficient plants. The traditional interpretation of potassium and calcium deficiency symptoms is that the high mobility of potassium ions within the plant allows the element to be continually recycled from the older to the developing leaves – so that the rate of new leaf production is maintained at the expense of the old. In contrast, calcium is relatively immobile in the plant and is retained in the leaves that receive it first. Consequently it is the birth rate of new leaves that is sacrificed.

One consequence of the different birth and death rates in normal and nutrient deficient plants is that the age structure of their foliage is dramatically altered. Figs.

Fig. 6.9. The changing age structure of the leaf population on plants of *Linum usitatissimum* grown in sand culture receiving (a) full Long Ashton nutrient solution, (b) nutrient solution omitting potassium, and (c) nutrient solution omitting calcium.

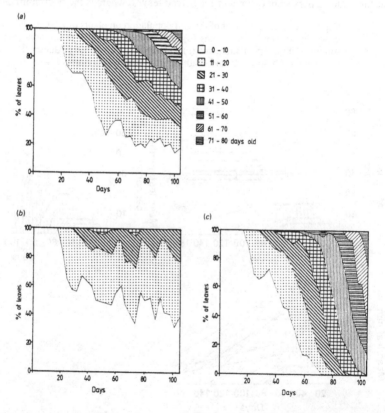

6.9(a,b) and (c) show the age structure of foliage as it develops over time. The plants receiving full nutrient solution continued to add new cohorts of leaves as they grew so that after 100 days the plants bore almost equal numbers of leaves in every age category from 0–9 days old to 70–80 days old. At the same time the calcium deficient plants bore no leaves younger than 40 days old and 20% of the foliage was more than 70 days old. The potassium deficient plants on the 100th day bore no leaves more than 30 days old and a quarter of the foliage was less than 9 days old. A full account of this experiment is given in Harper & Sellek, 1987.

The demography of leaves in some natural populations

Most studies of the birth and death of leaves have been made in managed agricultural crops and forests or on isolated plants. Two detailed studies have been

Fig. 6.10. (*a*) The birth rates per month of leaves of *Ammophila arenaria* recorded at 4–8 week intervals in plots of 1 m^2 along two transects across sand dunes at Newborough Warren, Anglesey, N.Wales. The transects crossed in turn, the first dune ridge, the young slack with impeded drainage, the second dune ridge, a fixed slack, dune grassland, an old grey dune and an old and now dry slack. (*b*) Death rates. These are shown only for leaves that had completed their life span during the study which accounts for the differences in scale in (*a*) and (*b*) (redrawn from Huiskes, 1977).

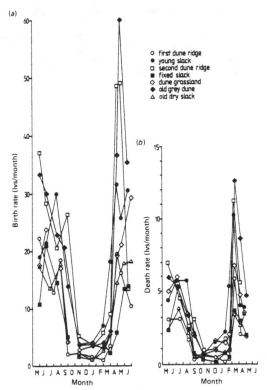

made on natural populations of the sand dune perennials *Ammophila arenaria* and *Carex arenaria*. Both studies involved deliberate perturbation of the natural communities and, in both, the unit of study was a defined area of vegetation rather than the individual plant.

Huiskes & Harper (1979) followed the birth and death of leaves of *Ammophila arenaria* in 1m² quadrats along transects across the sand dune system at Newborough Warren, Anglesey in the UK. The dramatic flux that underlies the apparent stability of the leaf populations is only exposed when cohorts of newly emerging leaves or shoots are marked and followed to their death. Birth rates and death rates are shown as seasonal cycles in Fig. 6.10. Most remarkable is the synchrony of births and deaths. As in other experiments reported in this paper, it is tempting to seek some causal interdependence for the two processes – leaf death creating the conditions (or releasing the resources) that allow new leaves to be produced, or leaf birth in some way forcing the death of old leaves. There is, however, no apparent time delay between the two processes that might hint at cause and effect.

The glasshouse experiments with *Lolium perenne* and field studies with *Trifolium repens* (Peters, 1980) showed that summer leaves had a short expectation of life and that winter-born leaves lived for a long time. The same effect is apparent in the life expectancy of leaves of *Ammophila* (Fig. 6.11) – leaves born in April and May lived for 5–10 weeks but those born in October to January remained alive for 15–25 (even

Fig. 6.11. The mean life expectancy at birth (in weeks) of leaves of *Ammophila arenaria* at different times of the year. The vertical bars show the 95% confidence limits. The columns show the means of the values calculated for each successional stage at different times of the year (from Huiskes, 1977).

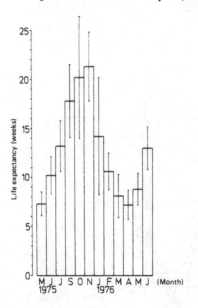

35) weeks in spite of, or perhaps because of, the slower growth rates in the winter. A crude interpretation (which does not actually say much) is that the life of a leaf is prolonged by a leisurely juvenile metabolism and curtailed by a vigorous summer life style.

The effect of nutrition on the dynamics of natural populations is shown clearly in studies of *Carex arenaria* on the Anglesey dunes. Noble *et al.* (1979) marked the emerging tillers of this rhizomatous sedge in quadrats placed at various positions in the sand dune zonation. This study took the whole tiller as the unit of demographic concern rather than the individual leaf and compared populations in control and fertilised plots (20g N, 8g P_2O_5, 14g K_2O_5 m^{-2}) applied in April and again in May before the study was started.

The effect of adding fertiliser was to speed up dramatically the rate at which the older cohorts died. Births and deaths in marked cohorts are shown (on a linear scale) together with the resulting changes in the total populations present (Fig. 6.12). The

Fig. 6.12. The dynamics of shoot cohorts of *Carex arenaria* born at successive time intervals in (*a*) control, and (*b*) fertilised plots in mature, senile and slack phases of dune vegetation. Arrows show the time of fertiliser application. Birth, life span and death of the separate shoot cohorts are shown with total shoot density superimposed (from Noble, Bell & Harper, 1979).

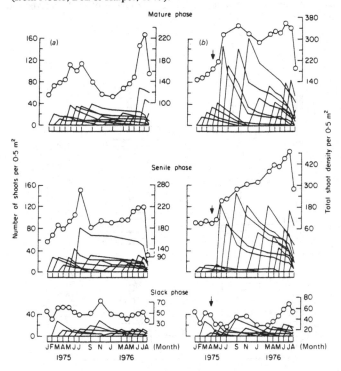

application of fertiliser clearly stimulated the birth of new shoots, especially in the mature and senile phases of the dune system.

A striking result of this experiment was the way in which the application of fertiliser increased the turnover rate of tillers (the flux) within the populations, despite the rather small changes in total population size. Once again, birth and death rates were tightly coupled. During the 20 months of this experiment, the fertilised population in the mature phase of the dunes increased from *ca.* 100 to *ca.* 200 shoots per 0.5 m^{-2}. This increase was the resultant of 1000 new births and 900 deaths within the quadrat! The effect was to convert an ageing population into one dominated by young shoots (Fig. 6.13).

The age structure of the foliage of trees

The pattern of development of a population of leaves on a tree is, at least in temperate species, very different from that of a typical annual plant such as *Linum*. The main difference is in the timing of births. In many deciduous and evergreen trees, shrubs and some perennial herbs, new leaves are expanded as an almost synchronous flush from preformed primordia in the bud. In these cases, the canopy can be regarded

Fig. 6.13. The age structure of shoot populations of *Carex arenaria* in control and fertilised plots in the mature, senile and slack phases of sand dune vegetation (from Noble, Bell & Harper, 1979).

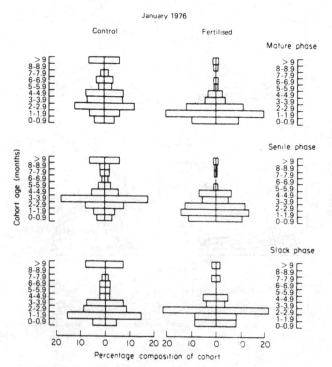

as a single cohort; pines are particularly good examples among evergreen trees in which one cohort of leaves is developed in a very short time each year and the leaves may persist for several years so that the canopy is composed of a set of annual cohorts, each very sharply separated in age from the next by 11–12 months.

In deciduous trees such as birch or willow, the first cohort of leaves again represents the expansion of those preformed in the bud, but a second phase follows in which further leaves are added (sometimes of a different shape) which develop from primordia formed in the same season (neoformed leaves).

In the deciduous trees most, but not all, leaves die synchronously in the autumn. In the tropics and in arid regions the demography may be more complex with rather sudden flushes of new leaf cohorts added to the canopy at intervals through the year. In others, for example *Teucrium polium*, seasonal cohorts of different shaped leaves succeed each other, each cohort persisting for only part of the year (Orshan, 1963).

The canopy dynamics of birch and of Corsican Pine (*Pinus nigra*) have been

Fig. 6.14. The age structure of needles on the canopy of *Pinus nigra* (from Maillette, 1979).

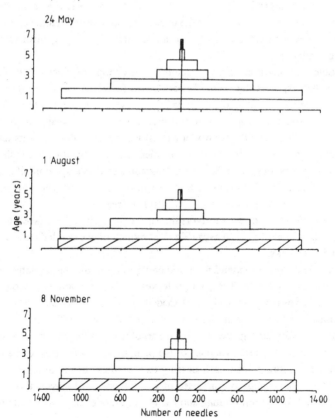

studied in some detail by Maillette (1982a,b and c) and illustrate the two main ways in which life tables can be constructed for demographic analysis. Maillette studied the canopies of three trees of *Pinus nigra* at the Botanic Garden, UCNW to determine their age structure. In pines it is easy to determine the age of branches from their morphology and so to determine the ages of the leaves that they bear. It is therefore possible *at one point in time* to determine, for every leaf on a tree, the year in which it was born. Because the leaves are expanded in one swift flush in the spring their ages are known with even greater precision.

Fig. 6.14 shows the characteristic age structure of the canopy as it changes in a single year. Data obtained in this way form a static life table. This can be very useful if, for example, we know that physiological changes occur in the leaves as they age and we wish to know something about the physiological responses of the canopy. It is a potentially misleading form of life table if we are concerned to analyse the demographic processes that have produced the age distribution. The tree in Fig. 6.14 bore many young and few old leaves. This situation could have arisen *either* if all leaves lived for 5 years and the tree had produced an increasing number of leaves each year *or* the tree had produced the same number of leaves each year but they had suffered a continuing risk of death. In the trees that Maillette studied the age structure of the canopy was the resultant of both an annually increasing birth rate and a continuing death risk.

The problem of interpreting age structure in the canopy of pines can be resolved because in these species the scars of dead leaves (needles) leave a record of past deaths. It is therefore possible to determine in this case, at one point in time, the shape of the survivorship curve for the cohort of leaves born in each year (Fig. 6.15). The way in which a pine bears the history of its past leaf deaths recorded in its leaf scars means that it is possible to determine all the elements in a dynamic life table from a study made at one point in time. In most demographic studies a dynamic life table can only be constructed by following each cohort through to the death of its members as in most of the other studies described in this paper.

In the pines the shape of the survivorship curves is clearly Type 1 (see Fig. 6.1) of the same form as that of grass leaves (Fig. 6.2) though now extended over 6–7 years instead of 5–14 weeks.

Because trees grow by extending a branching system, the age structure of their canopy is not homogeneous. The younger leaves are born in more exposed positions and the older leaves are more likely to be shaded by the young. This is particularly the case in pines in which it would be possible to draw contours through the canopy representing both different age structures of the needles and different degrees of shading.

In the analysis of the leaf dynamics of the pine, needles were regarded as leaves: strictly they are short shoots, each bearing two needles. In the birch, as in many deciduous trees, the canopy is composed of two categories of leaves: those born on short shoots and those born on long shoots. The internodes of short shoots scarcely elongate during the year and bear a rosette or 1–6 leaves. The internodes of long shoots elongate during the season and may bear 60 or more leaves. The leaves born on the long shoots include those that were preformed in the bud in the previous year, and to some extent reflect the conditions in that year, and also those that were initiated as primordia and continued to complete their development and expansion in the current year.

Short shoots are concentrated on the older branches in the more shaded parts of the canopy, whereas the long shoots form leaders, extending the fringes of the canopy, particularly at the top (Fig. 6.16).

Fig. 6.17 shows the annual cycle of the leaf population on three free-standing birch trees and the contribution made to this by the short and long shoots. Note that in every case the maximum number of leaves present fell well short of the number actually produced. Clearly, mortality had been a significant force in determining the

Fig. 6.15. The survivorship curves of needles of *Pinus nigra* on first-, second-, third- and fourth-order branches (from Maillette, 1979).

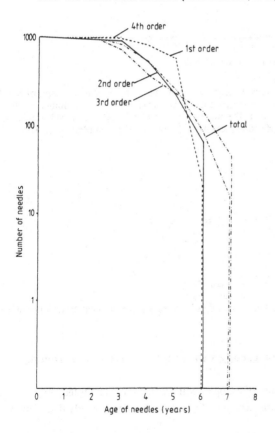

Fig. 6.16. The frequency distribution of leaves born on short and long shoots of *Betula pendula* in the top of the canopy and over the whole tree (from Maillette, 1979).

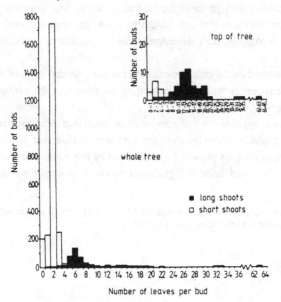

Fig. 6.17. The annual cycle of the leaf population on free-standing trees of *Betula pendula* (mean value for three trees). Leaves from short (---) and long (...) shoots have been distinguished. The maximum number of leaves produced was 20 000 (total), 11 000 (short shoots) and 9 000 (long shoots) (from Maillette, 1979).

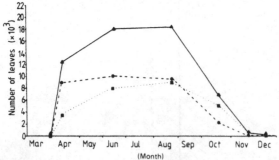

composition of the canopy. The risk of death appears to be greatest among the leaves on short shoots.

The relevance of canopy demography to ecology, agronomy and forestry

Demographic studies of canopies will become important if they improve our predictions of canopy performance. In traditional demography, largely that of humans and other animals, the study of birth and death rates is made to predict future and explain past changes in population growth rates and, especially for humans, to prepare actuarial tables for risk assessment. Simple counts of the numbers or mass in a population have no predictive value. Clearly, the number of offspring left by the inhabitants of an old people's home, a school and a new population of immigrants will not be the same. In the same way, the future photosynthetic rate of a canopy will not be independent of its age structure.

To predict the growth rate of a population of animals we can combine knowledge of the number of individuals of different ages with information about the number of offspring produced by individuals of different ages. The life table that is used for this purpose includes the age-specific probabilities of survival for the individuals. The life table is combined with a fecundity schedule which lists age-specific reproductive rates in parallel with the life table. In any population the probability of an individual of age x producing offspring over the short element of age dx is $l_x t_x \, dx$ where l_x is derived from the survivorship curve and b_x from the fecundity curve. This expression can be integrated to give the total expectation of offspring. If the age and specific fecundity distributions remain constant the rate of increase of the population is also constant. The essence of this procedure is that it should predict future populations' growth from present observations.

It is not difficult to see the parallels between predicting the growth of a population of animals and predicting the activities of a population of leaves. Instead of a fecundity schedule we need a schedule that specifies the way in which some activity, for example photosynthesis or net assimilation rate, changes with leaf age. If we could combine this with knowledge of the age structure of the leaf population we would have most of the elements needed for a predictive theory of canopy performance. In practice we cannot yet do this because we know the age structures and age-specific activity schedules for very few species and far fewer genotypes. Moreover, the activity of a leaf is determined not only by its age but also by its position in the canopy. In practice, as has already been pointed out, leaf age usually determines the position occupied in the canopy so it may be possible to treat position in the canopy as just one of the age-related characteristics of a leaf's performance.

Fig. 6.18. Changes in the nutrient contents of the leaves of *Salix cinerea* with age (from Alliende, 1986).

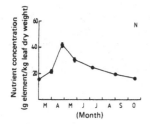 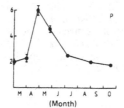 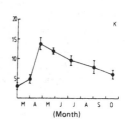

A classic example of an age-specific activity curve is that for the leaf of the cucumber (*Cucumis sativus*) (Hopkinson, 1964) which was deliberately chosen for study because its leaves can be arranged in an experiment so that they do not overlap each other. This is precisely the type of data that might be taken together with information on leaf birth rates and survivorship to build a full demographic model to predict canopy performance.

There are other important ways in which demographic analysis may contribute to understanding the ways in which canopies behave. A leaf is not just an organ of carbon assimilation – it is also a region of nutrient accumulation and subsequent dissipation (see van Keulen, Goudriaan & Seligman in this volume). Differences between species in the longevity of leaves may be related to their role as sites of mineral storage and it may be relevant that the evergreen habit seems often to be characteristic of species living in environments in which mineral resources are scarce or rapidly leached. Thus the demography of foliage may provide clues to understanding nutrient cycling (Moore, 1984).

A leaf may contribute to the success of a plant by shading potential competitors. Even an old leaf (or a dead one) may play a part in the success of one plant overtopping a neighbour if it deprives the neighbour of light.

The age structure of a canopy determines its value as food for herbivores. There is much evidence that many folivorous insects discriminate between leaves on the basis of nutritive value, especially their nitrogen content (Crawley, 1983), which is age-dependent (Fig. 6.18). The feeding value of a grazed sward to farm livestock changes dramatically with age, so much so that it is normal agricultural practice to sacrifice the bulk of a mature hay or silage crop in favour of the less bulky, but nutritively more valuable, juvenile foliage.

This chapter gives only an overview of the subject of canopies as populations. The bibliography includes a number of references that will lead an interested reader deeper into the literature.

References

Alliende, M.C. (1986). *Growth and Reproduction in a Dioecious Tree*, Salix cinerea. PhD thesis, University of Wales.

Baker, C.K., Gallagher, J.N. & Monteith, J.L. (1980). Daylength change and leaf appearance in winter wheat. *Plant Cell and Environment*, **3**, 285–7.

Bazzaz, F.A. & Harper, J.L. (1977). Demographic analysis of the growth of *Linum usitatissimum. New Phytologist*, **78**, 193–208.

Bonner, F.T. & Dell, T.R. (1976). The Weibull function: a new method of comparing seed vigor. *Journal of Seed Technology*, **1**, 96–103.

Clark, S.C. (1980). Reproductive and vegetative performance in two winter annual grasses, *Catapodium rigidum* (*L.*) *C.E. Hubbard* and *C. murinum* (*L.*) *C.E. Hubbard*. II. Leaf demography and its relationship to the production of caryopses. *New Phytologist*, **84**, 79–93.

Crawley, M.J. (1983). *Herbivory: The Dynamics of Animal–Plant Interactions.* Oxford: Blackwell.

Ellis, R.P. & Russell, G. (1984). Plant development and grain yield in spring and winter barley. *Journal of Agricultural Science, Cambridge*, **102**, 85–95.

Gallagher, J.N. (1979). Field studies of cereal leaf growth. I. Initiation and expansion in relation to temperature and ontogeny. *Journal of Experimental Botany*, **30**, 625–36.

Gehan, E.A. & Siddiqui, M.M. (1973). Simple regression methods for survival time studies. *Journal of the American Statistical Association*, **68**, 848–56.

Harper, J.L. & Selleck, C. (1987). The effects of severe mineral nutrient deficiencies on the demography of leaves. *Proceedings of the Royal Society*, **B232**, 137–57.

Hewitt, E. (1966). Sand and water culture methods used in the study of plant nutrition. *Commonwealth Bureau of Horticulture and Plantation Crops, East Malling, Technical Communication*, **22**, 1–547.

Hopkinson, J.M. (1964). Studies on the expansion of the leaf surface. IV. The carbon and phosphorus economy of a leaf. *Journal of Experimental Botany*, **15**, 125–37.

Huiskes, A.H.L. (1977). *The Population Dynamics of Carex arenaria (L.) Link.* PhD thesis, University of Wales.

Huiskes, A.H.L. & Harper, J.L. (1979). The demography of leaves and tillers of *Ammophila arenaria* in a dune sere. *Oecologia Plantarum*, **14**, 435–46.

Kirby, E.J.M., Appleyard, M. & Fellowes, G. (1982). Effect of sowing date on the temperature response of leaf emergence and leaf size in barley. *Plant Cell and Environment*, **5**, 477–84.

Klepper, B., Rickman, R.W. & Peterson, C.M. (1982). Quantitative characterisation of vegetative development in small grain cereals. *Agronomy Journal*, **74**, 789–92.

Maillette, L. (1979). *Structural Dynamics of Tree Growth, with Special Reference to* Betula pendula Roth. *and* Pinus nigra var maritima *(Ait.) Melville. PhD thesis. University of Wales.*

Maillette, L. (1982a). Needle demography and growth dynamics of Corsican Pine. *Canadian Journal of Botany*, **60**, 105–16.

Maillette, L. (1982b). Structural dynamics of silver birch. I. The fates of buds. *Journal of Applied Ecology*, **19**, 203–18.

Maillette, L. (1982c). Structural dynamics of silver birch. II. A matrix model of the bud population. *Journal of Applied Ecology*, **19**, 219–38.

Masle–Meynard, J. & Sebillotte, M. (1981). Etude de l'hétérogénéite d'un peuplement de blé d'hiver. Notion de structure du peuplement. *Agronomie*, **1**, 207–16.

Moore, P.D. (1984). Why be an evergreen? *Nature (London)*, **312**, 703.

Noble, J.C., Bell, A.D. & Harper, J.L. (1979). The population biology of plants with clonal growth. I. The morphology and structural demography of *Carex arenaria*. *Journal of Ecology*, **67**, 983–1008.

Orshan, G. (1963). Seasonal dimorphism of desert and mediterranean chamaephytes and its significance as a factor in their water economy. In *The Water Relations of Plants*, ed. A.J. Rutter & F.H. Whitehead, pp. 207–22. Third Symposium of the British Ecological Society. Oxford: Blackwell.

Peters, B. (1980). *The Demography of Leaves in a Permanent Pasture. PhD thesis, University of Wales* .

Pinder, J.E., Wiener, J.G. & Smith, M.H. (1978). The Weibull distribution: A new method of summarising survivorship data. *Ecology*, **59**, 175–9.

Pyke, D.A. & Thompson, J.N. (1986). Statistical analysis of survival and removal rate experiments. *Ecology*, **67**, 240–5.

128   J. L. HARPER

Wallace, T. (1944). *The Diagnosis of Mineral Deficiencies in Plants by Visual Symptoms: a Colour Atlas and Guide.* London: HMSO.

Williamson, P. (1976). Above ground primary production of chalk grassland allowing for leaf death. *Journal of Ecology,* **64**, 1059–75.

JAMES R. EHLERINGER AND
IRWIN N. FORSETH

7. Diurnal leaf movements and productivity in canopies

Introduction

Over the past 15 years a number of studies have focused on characterising diurnal leaf movements that occur in a variety of plants in response to the sun's movement across the sky. It is now clear that these solar tracking leaf movements are triggered by a directional light stimulus and that these movements result in at least a partial regulation by the leaf of the intensity of the incident photon irradiance. The purpose of this chapter is to review what is known about the different kinds of leaf solar tracking movements, their impact on primary productivity, and the potential ecological roles of these phenomena.

Solar tracking is an expression applied to describe the heliotropic movements of both leaves and flowers; it denotes the ability of these structures to move in response to the diurnal change in the sun's position in the sky. Heliotropic movements are distinguishable from other directional types of growth by their rapidity, the reversibility and by the overnight resetting to face the morning sun (Yin, 1938). Two main kinds of diurnal movements are recognised: diaheliotropic movements in which the leaf lamina remain oriented perpendicular to the sun's direct rays and paraheliotropic movements in which the leaf lamina are oriented obliquely to the sun's direct rays (Ehleringer & Forseth, 1980). In the extreme cases of paraheliotropism, the leaf lamina may change from nearly perpendicular to the sun's rays to an orientation parallel to the sun's rays.

The first and most immediate consequence of leaf solar tracking is to regulate the level of the incident photon irradiance from the sun. This means that solar tracking movements can be used to maximise or minimise photon irradiance incident on the leaf (Fig. 7.1). Since only the direct component of photon irradiance can be regulated, the minimum incident photon irradiance in Fig. 7.1 represents the incident component of diffuse solar radiation. Through such leaf movements, exposed canopy leaves are not only able to regulate the intensity of the peak photon irradiance, but also its diurnal timing and the daily receipt of photons. Fig. 7.1 shows that, depending on the water stress levels imposed on the plant (as will be discussed later), the peak photon irradiances on a paraheliotropic leaf (*a,b* or *c*) can occur at any time between early morning and late afternoon.

Solar tracking movements may be quantified by calculating the cosine of the angle between the normal to the leaf lamina and the direct solar beam, i.e., the angle of incidence. The cosine varies between 0 and 1 depending on the relative geometrical positions of the leaf lamina and the sun, and is calculated as

$$\cos{(i)} = \cos(\alpha_l)\sin(\alpha_s) + \sin(\alpha_l)\cos(\alpha_s)\cos(\beta_s-\beta_l) \qquad (1)$$

where α_l is the angle above the horizontal of the long axis of the leaf lamina, α_s is the angle of the sun from the horizontal and β_l and β_s are the azimuthal positions of the leaf and sun respectively (Gates, 1962). A value of $\cos(i)$ close to 1.0 throughout the day indicates a leaf with a strong diaheliotropic ability. If the leaves in a canopy do not move and have a random leaf azimuth distribution, the average $\cos(i)$ should vary sinusoidally through the day as the sun passes across the sky.

The mechanism responsible for solar tracking movements appears to be turgor-mediated and is similar to that involved in other types of rapid leaf movements such as seismonasty and nyctinasty (Schwartz & Koller, 1978; Satter & Galston, 1981; Vogelmann & Björn, 1983). However, the receptor sites that detect direct sunlight appear to differ between species. For example, in two species from the Malvaceae (*Malva neglecta* and *Lavatera cretica*) orientation is controlled by turgor changes in a pulvinus located at the junction of the lamina and the petiole, while the light perception sites appear to be in the major veins of the lamina (Yin, 1938; Schwartz & Koller, 1978). In contrast, *Lupinus succulentus* from the Fabaceae exhibits both light perception and turgor changes in the pulvinus (Vogelmann, 1984). One hypothesis advanced to explain the first type of sensing mechanism involves differential carbon assimilation patterns of mesophyll cells on either side of major veins due to

Fig. 7.1. The diurnal courses of incident photon irradiance (400–700 nm) on a diaheliotropic leaf ($\cos(i)=1$), two paraheliotropic leaves exhibiting diurnal decreases in $\cos(i)$, a paraheliotropic leaf, ($\cos(i)=0.1$), and a horizontal leaf. Based on Shackel & Hall (1979) and Ehleringer & Forseth (1980).

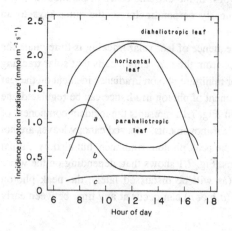

microtopographical shading effects. This differential production of carbohydrates would then result in the transmission of some signal to the cells of the pulvinus, initiating reorientation (Fisher & Fisher, 1983). The tracking photoreceptor in both types is activated by blue light (Yin, 1938; Koller, 1981; Vogelmann & Björn, 1983; Sheriff & Ludlow, 1985), making it more effective for fully sunlit leaves than for those within the canopy.

There is currently little information as to the extent of solar tracking among higher plants, but those data that are available clearly suggest that it is frequent among the annual and herbaceous vegetation of arid and semi-arid habitats (Begg & Torssell, 1974; Wainwright, 1977; Ehleringer & Forseth, 1980; Sheriff & Ludlow, 1985). Solar tracking also occurs in plants from temperate and sub-tropical habitats, especially among legumes and other compound leaved species (Gates, 1916; Herbert, 1984; Herbert & Larsen, 1985; Forseth & Teramura, 1986). Its occurrence among crop plants has been noted for many species, including bean (*Phaseolus vulgaris*) (Dubetz, 1969; Wien & Wallace, 1973), cotton (*Gossypium hirsutum*) (Lang, 1973; Ehleringer & Hammond, 1987), soybean (*Glycine max*) (Meyer & Walker, 1981; Wofford & Allen, 1982; Oosterhuis, Walker & Eastham, 1985), sunflower (*Helianthus annuus*) (Shell, Lang & Sale, 1974; Shell & Lang, 1976; Lang & Begg, 1979) and alfalfa (*Medicago sativa*) (Travis & Reed, 1983). Undoubtedly, many more examples exist and our biogeographic understanding of this phenomenon is far from complete.

Solar tracking and photosynthesis

The maintenance of a high cosine of incidence throughout the day in diaheliotropic leaves will result in relatively constant high photon irradiances (near 2.0 mmol m^{-2} s^{-1}, 400–700 nm). For paraheliotropic leaves, the incident photon irradiances will also remain nearly constant but at a reduced level. What are the photosynthetic characteristics of such leaves?

A small number of studies have been conducted on plants with diaheliotropic leaves grown under the high-irradiance conditions typical of their natural environments. These data indicate that rapid photosynthetic rates can be expected under midday light levels (Fig. 7.2). Photosynthetic rates exceeding 50 μmol m^{-2} s^{-1} have been recorded for a number of native desert species in bright light (Toft & Pearcy, 1982; Forseth & Ehleringer, 1983a; Werk *et al.*, 1983). These rates are achieved not by having an intrinsically higher quantum yield, but rather by being able to utilise bright light more efficiently. That is, the only difference between the photosynthetic characteristics of these plants and those with lower photosynthetic rates is that they do not light saturate under the conditions they normally experience.

Not all plants with diaheliotropic leaf movements exhibit such high photosynthetic rates. Two common crop species, sunflower and cotton, have maximum photosynthetic rates betwen 25 and 35 μmol m^{-2} s^{-1} (Fig. 2). While these rates are indeed high compared with other C_3 species (Black, 1973; Bazzaz, 1979) they are substantially lower than in native plants that exhibit solar tracking. Photosynthesis in both sunflower and cotton appears to be light saturated at irradiances less than those typically incident on the leaf through the day. In fact, Ehleringer & Hammond (1987) demonstrated that solar tracking cotton leaves were light saturated by a photon irradiance of 1.2 mmol m^{-2} s^{-1}, so that during the day leaves were experiencing irradiances 40% beyond what they were able to utilise in photosynthesis. The significance of a reduced photosynthetic rate and photosynthetic light utilisation in these crops is unclear. It is, however, important to note that both of these species have undergone significant selection during breeding so that some aspects of current plant performance may differ from those found in more primitive forms. In addition, solar tracking movements do not occur in all cultivated cotton varieties (Ehleringer & Hammond, 1987) and leaf movements in domesticated sunflowers decrease through time in mature plants (Lang & Begg, 1979).

In both native and crop species with paraheliotropic leaf movements, the gas exchange data indicate that leaves are operating near the upper end of the linear portion of the photosynthesis–light response curve (Fig. 7.2; *Lupinus, Pueraria, Glycine* and *Medicago*). This indicates that while the absolute photosynthetic rate of paraheliotropic leaves may be lower than that of diaheliotropic leaves, the efficiency of utilisation of incident light appears to be higher. The arrows in Fig. 7.2 indicate the average midday photon irradiance incident on the leaves. In all cases, it appears that paraheliotropism has resulted in a modification of the incident light so that leaves are operating over a

Fig. 7.2. The dependence of net CO_2 uptake by leaves of species exhibiting solar-tracking leaf movements. Based on Werk *et al.* (1983), Travis & Reed (1983), Ehleringer & Hammond (1987), and I. Forseth (unpublished observations).

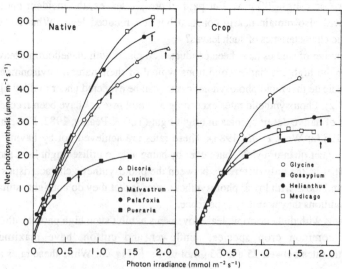

portion of the response curve where changes in incident photon irradiance have a pronounced effect on photosynthetic rates.

Diurnal leaf movements have two principal effects on environmental aspects of photosynthesis: (1) they provide a mechanism whereby the plant is able to achieve favourable photosynthetic rates at specific times during the day, and (2) they allow the leaf to avoid high incident photon irradiances at times of the day that are not favourable for photosynthesis. In particular, since stomata are sensitive to humidity and close under reduced humidities (Schulze & Hall, 1982), solar tracking movements result in higher incident light levels, and thus faster photosynthetic rates, early in the morning and late in the afternoon when humidity levels are typically highest (Bonhomme, Varlet Grancher & Artis, 1974; Mooney & Ehleringer, 1978; Forseth & Ehleringer, 1983b). When the availability of soil water is limited, paraheliotropic leaves appear to have an advantage over strictly diaheliotropic leaves because they are able to vary the fraction of the direct solar beam that is incident. This allows the leaves to capitalise on early morning sunlight and to minimise midday and afternoon sunlight if soil water conditions are unfavourable (Shackel & Hall, 1979). In general, it appears that paraheliotropic leaves move diurnally to adjust incident irradiances so that photosynthetic rates are not light saturated (Shackel & Hall, 1979; Travis & Reed, 1983; Forseth & Teramura, 1986).

Leaf movements to regulate incident solar radiation

While paraheliotropic leaf movements regulate the diurnal patterns of incident light, the significance of this phenomenon depends on the availability of soil water. Under conditions of adequate soil water, plants with the potential for paraheliotropic movements appear to be reacting to short-term midday conditions of high evaporative demand in such a way as to regulate light interception, so that a balance is struck between light level and gas exchange activity.

Some native and cultivated leguminous species growing in conditions of adequate soil water show constant diurnal values of $\cos(i)$ (Kawashima, 1969a, b; Forseth & Ehleringer, 1980; Forseth & Teramura, 1986). The value of this constant $\cos(i)$ varies between 0.4 and 0.6 and results in little variation in incident irradiance for these species throughout the day. This regulation of incident radiation will have several effects. First, since transpiration is likely to increase with increasing incident irradiance while photosynthesis shows a saturating response, water use efficiency will tend to be enhanced (Rawson, 1979). Radiation not necessary for photosynthetic carbon gain is thus not intercepted, reducing heat loads. Secondly, the maintenance of this constant $\cos(i)$ by outer canopy leaves will allow much greater light penetration into the lower canopy. This may enhance the contribution of leaves lower in the canopy to the overall carbon fixation of the canopy (Travis & Reed, 1983). Finally, the coincidence of peak irradiance levels and evaporative demand at midday experienced by horizontally displayed leaves is not experienced by these paraheliotropic leaves (Fig. 7.4(*b*)). This results in the avoidance of short-term water stress conditions that may be induced by high transpiration rates even under conditions of high soil-water availability.

There are numerous reports of plant species being able to adjust leaf angles in response to limited soil moisture conditions (Dubetz, 1969; Wien & Wallace, 1973; Begg & Torssell, 1974; Shackel & Hall, 1979; Forseth & Ehleringer, 1980; Meyer & Walker, 1981; Herbert, 1984; Oosterhuis *et al.*, 1985). These studies indicate that under soil water stress, the leaves or leaflets increase their leaf angle such that they are near vertical at midday. Most of these species have compound leaves, and the "cupping movements" that result in these steep leaf angles involve leaflet folding

Fig. 7.3. Leaflet orientations in *Lupinus arizonicus* under well-watered (left), and droughted (right), soil conditions with the sun's position being in the upper left.

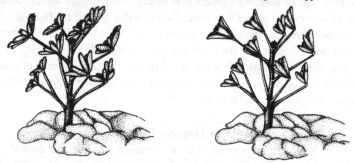

Fig. 7.4.(*a*). The dependence of the cosine of the angle of incidence for leaves of *Lupinus arizonicus* as a function of leaf water potential. Based on Forseth & Ehleringer (1980), and (*b*) the calculated solar radiation levels incident on leaflets of *Lupinus arizonicus* through the course of the day on 1 April (typical midpoint of growing season) as a function of the cosine of the angle of incidence. Based on Forseth & Ehleringer (1983*b*).

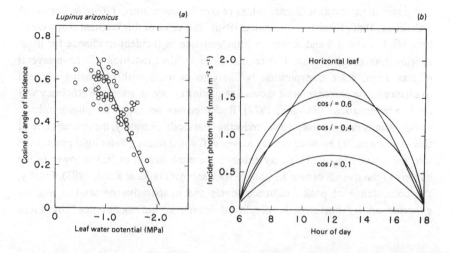

around the basal pulvinal region near the tip of the petiole (Fig. 7.3). While most investigators have focused their observations on the midday period, diurnal measurements of cos(i) values on water-stressed plants indicate that these sun-avoidance movements occur throughout the day (Shackel & Hall, 1979; Forseth & Ehleringer, 1980, 1983a).

Further, instead of being an all-or-none phenomenon, cos(i) varies continuously with leaf water stress (Begg & Torssell, 1974; Shackel & Hall, 1979; Forseth & Ehleringer, 1980, 1983a). The response appears to be plastic and is rapidly reversible when the stress is alleviated. In *Lupinus arizonicus*, leaf cos(i) is closely related to leaf water potential (Fig. 7.4(a)). The responses of cos(i) to water stress for different field populations or for greenhouse versus field-grown materials are indistinguishable. There appears to be a tight control over leaf movements in *L. arizonicus*, so that neither leaf cos(i) nor leaf water potential exhibit much diurnal fluctuation (Forseth & Ehleringer, 1982). This results in less variable light levels on the leaf through the day (Fig. 7.4(b)). Local microsite variations in soil depth and water content in these desert soils, however, result in leaves from different plants within a single population exhibiting different values of cos(i). Since both leaf orientation and leaf conductance to water vapour are tightly coupled to the leaf water potential (Forseth & Ehleringer, 1982), the consequence is that adjacent plants can be operating quite differently depending on local soil conditions.

Under dense canopies when soil water availability is low, paraheliotropic movements will have dramatic effects upon canopy microclimate. In a planophile canopy, radiation absorption, and hence leaf temperature maxima, is concentrated in the upper few leaf layers. The vertical leaf angles achieved by water-stressed soybean crops (Meyer & Walker, 1981; Oosterhuis *et al.*, 1985) have the effect of displacing these maxima towards the ground surface (Baldocchi, Verma & Rosenberg, 1983). Since water stress significantly reduces photosythetic capacity and the ability of plants to use bright light for carbon dioxide uptake (Hsiao, 1973; Bradford & Hsiao, 1982; Forseth & Ehleringer, 1983(a)) the reduction in the amount of radiation intercepted by the canopy will have minimal impact on productivity. Also, due to these changes in leaf angle, radiation extinction coefficients will show marked diurnal variation, with a midday minimum and morning and afternoon maxima (Baldocchi *et al.*, 1983). Depending upon soil water status, paraheliotropic crops with high leaf area indices (LAI) such as soybean and cowpea (*Vigna unguiculata*) are thus able to modify radiation interception between that of a planophile canopy with high, constant extinction coefficients to that of an erectophile canopy with diurnal changes in extinction coefficients (Shackel & Hall, 1979; Baldocchi *et al.*, 1983).

There are four ways in which net plant performance benefits from paraheliotropic leaf movements under conditions of water stress. Reducing the amount of light incident upon the leaf greatly modifies the leaf energy balance through a reduction in the incident heat load. This has two beneficial effects: a lowered leaf temperature and a consequent reduction in transpiration rate without necessarily any change in leaf conductance. Temperature changes can be substantial. Forseth & Teramura (1986) demonstrated that leaf temperatures of the weedy vine *Pueraria lobata* increased by 4 to 5 °C when leaves were prevented from cupping (Fig. 7.5). Such increases in leaf temperature will cause the transpiration rate to increase by 60–80%, solely because of the temperature-associated changes in the leaf to air water vapour gradient. A third benefit is that leaflet reorientation results in an increased water use efficiency of the leaves (photosynthesis/transpiration ratio). Under water-limited conditions when leaf conductances are already reduced, a reduction in cos(i) has a greater effect on lowering leaf temperature (and thus transpiration) than it does on reducing photosynthesis (Forseth & Ehleringer, 1983(b)). Lastly, a reduction in the amount of light absorbed by the leaf under water stress conditions reduces the likelihood of photo-inhibitory damage. Ludlow & Björkman (1984) have shown that if the reversible leaf movements are prevented in *Macroptilium atropurpureum*, a plant

Fig. 7.5. Photon irradiance (400–700 nm) and air temperatures adjacent to kudzu (*Pueraria lobata*) leaves near College Park, Maryland on 21 July and the corresponding leaf temperatures of naturally orientated and horizontally restrained leaves of kudzu (based on Forseth & Teramura, 1986).

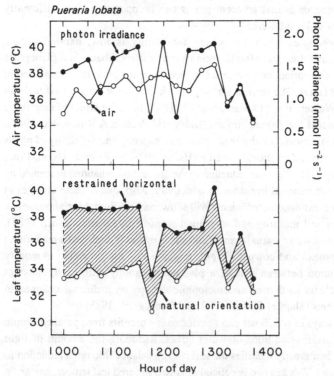

which typically exhibits paraheliotropic leaf movements under conditions of water stress, there is a time-dependent reduction in photosynthetic activity caused by exposure to bright light.

Limitations associated with diurnal leaf movements

Diaheliotropic leaf movements maximise the interception of direct solar irradiation. This has consequences for canopy structure, light penetration within canopies, and on canopy size.

For most native desert annuals in North America, individuals are widely spaced and LAI values are generally between 1.0 and 1.5 (Adams, Strain & Adams, 1970; Beatley, 1969; Ehleringer & Mooney, 1983). Thus, for those species with either diaheliotropic or paraheliotropic leaf movements, effectively all leaves are fully exposed to sunlight. Canopy development is therefore constrained more by the period of adequate soil moisture than it is by light levels within the canopy leaves.

By contrast, in intensive agriculture sufficient water is available to the plants for a longer period of time and so they support a larger leaf area. In row-planted cotton, the upper canopy leaves are fully illuminated and exhibit strong diaheliotropic movements (Lang, 1973; Ehleringer & Hammond, 1987). As shown in Fig. 7.6, almost all the upper canopy leaves in a stand with an LAI of approximately 4 were oriented to within 60° of the sun's azimuth both early in the morning and again late in the afternoon. In contrast, the leaves in the lower canopy did not exhibit any pronounced tendency to orientate towards the sun. In the early morning the distribution of these leaves is uniform with respect to azimuth. By the end of the day there was an indication that a significant fraction was orientated to the south, probably reflecting the higher light levels on the southern side of these crops, which were planted in east–west rows. Data on light penetration into cotton canopies, indicate a high light extinction coefficient within the top 30 cm associated with these strong diaheliotropic leaf movements (Fukai & Loomis, 1976). Since a significant direct solar beam is required to achieve solar tracking movements, few of the inner canopy leaves will receive sufficient radiation to maintain diurnal movements. The consequence is that the bulk of the canopy productivity is restricted to the leaves of the extreme upper canopy (Fukai & Loomis, 1976). However, from Fig. 7.2, it appears that these leaves are not able to capitalise on the elevated photon irradiances, because they are light saturated at only 1.0 mmol m^{-2} s^{-1}. Thus, solar tracking movements in cotton do not enhance overall canopy productivity, and may indeed reduce it. Such was the conclusion of Fukai & Loomis (1976), who modelled total canopy productivity in cotton. When LAI was between 1 and 2, leaf solar tracking enhanced canopy productivity since leaves were absorbing photons which would otherwise pass through the sparse canopy. However, when LAI was above 4, simulation results indicated that leaf solar tracking reduced canopy productivity. The conclusion was that total canopy production would have been increased if the uppermost canopy leaves had not absorbed such a significant fraction of the incident light, but instead had allowed for greater light penetration into the canopy.

Begg & Jarvis (1968) investigated photosynthetic characteristics and canopy productivity in Townsville stylo (*Stylosanthes humilis*), a plant with diaheliotropic movements under well-watered conditions and paraheliotropic movements under droughted conditions. Their observation was that while less than 10% of the total LAI was located in the top 10 cm, this layer absorbed 80–86% of the sunlight under well-watered conditions. Their photosynthetic canopy models predicted that the maximum canopy productivity would be attained with a 20 cm canopy, and that productivity should be lower in taller canopies.

In contrast to diaheliotropic movements, paraheliotropic leaf movements are not expected to limit canopy size. Instead, paraheliotropism under well-watered conditions provides a mechanism whereby outer canopy leaves can regulate the extent of light penetration into lower canopy levels. Travis & Reed (1983) explored the consequences of solar tracking on light penetration into alfalfa canopies. Photosynthetic rates in alfalfa are light saturated at 1 mmol m^{-2} s^{-1} (Fig. 7.2). In alfalfa, $\cos(i)$ increases with depth into the canopy. Thus, the outermost leaves have a low $\cos(i)$, allowing a greater light penetration into the canopy. Since the lower canopy leaves have a $\cos(i)$ near 1.0 a greater fraction of the leaves in the canopy are able to operate at maximal photosynthetic rates. In contrast to the Townsville stylo

Fig. 7.6. The distributions of leaf azimuths of a solar-tracking cotton (*Gossypium hirsutum* cv. Delta Pine 62) measured at the top and bottom of a canopy in the early morning and again in the late afternoon. Based on Ehleringer & Hammond (1987).

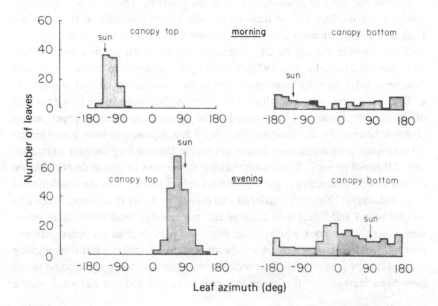

canopy, alfalfa canopies exhibit maximum productivity at a canopy height of 65 cm (Stanhill, 1962).

Ecological patterns

Leaf solar tracking movements only influence plant performance when the ratio of direct to diffuse components of the solar radiation is high since only in these conditions will leaf movements be able to regulate or affect the incident light levels. This implies that diaheliotropism and paraheliotropism should be restricted to environments receiving high daily total photon irradiances or environments in which a significant fraction of the day is clear, i.e., they would not be expected to occur in habitats with a high incidence of overcast days or in understorey habitats.

Diaheliotropic leaf movements should have their greatest benefit in arid lands and in habitats with strong monsoon activity because in these situations the canopy LAI is low due to soil moisture limitations or short growing seasons.

This is in fact the pattern that has emerged. By far the overwhelming majority of species reported as having diaheliotropic leaf movements are in desert environments with short, unpredictable growing seasons, or in weedy, disturbed habitats (Yin, 1938; Schwartz & Koller, 1978; Ehleringer & Forseth, 1980).

Paraheliotropic leaf movements can be expected in any environment with a high direct irradiance and where either diurnal or seasonal environmentally-induced stress may occur. It is uncertain whether high temperature, photosynthetic photoinhibition, high rates of transpiration, or soil-induced water stress is more important for its occurrence. Singly, or in combination they may all have played a role in the evolution of paraheliotropism. So far, reports of its occurrence range from 33 leguminous species in the seasonally-dry tropical environment of the Philippine Islands (Gates, 1916) to temperate tree species (McMillen & McClendon, 1979), subtropical legumes (Herbert, 1984), temperate lianas (Forseth & Teramura, 1986), leguminous crops (Dubetz, 1969; Wien & Wallace, 1973; Shackel & Hall, 1979; Meyer & Walker, 1981), semi-arid pasture legumes (Sheriff & Ludlow, 1985) and finally to arid land trees and annuals (Wainwright, 1977; Ehleringer & Forseth, 1980). These examples span a range of values of LAI from less than 1 to more than 5. This indicates that these leaf movements are not necessarily detrimental to productivity and may actually enhance productivity on a canopy basis.

Paraheliotropic movements are particularly evident during midday periods when the sun is high in the sky. Under these conditions, leaf angle changes affect $\cos(i)$ to a greater extent than do changes in leaf azimuth. In fact, most reports show random distributions of leaf azimuth during midday periods in paraheliotropic species (McMillen & McClendon, 1979; Travis & Reed, 1983; Oosterhuis *et al.*, 1985; Forseth & Teramura, 1986). Several lines of evidence, including (1) the requirement only for angle changes for midday effects, and (2) the widespread occurrence of paraheliotropism in many habitats as well as in situations of both high and low canopy development, suggest that this type of movement may have evolved first. Paraheliotropism is widespread in compound-leaved species, especially in the Fabaceae (Gates, 1916; McMillen & McClendon, 1979; Herbert & Larsen, 1985). Compound leaves may represent a preadaptation for rapid changes of leaf angle at midday. This feature may explain the dramatic increase in the number of canopy species with compound leaves as you move along an environmental gradient from wet, evergreen tropical forests to seasonally dry, deciduous tropical forests (Givnish, 1978). Further evolutionary and biogeographical musings regarding leaf solar tracking must, however, await a more complete survey of habitats.

References

Adams, S., Strain, B.R. & Adams, M.S. (1970). Water-repellent soils, fire and annual plant cover in a desert scrub community of southeastern California. *Ecology*, **51**, 696–700.

Baldocchi, D.B., Verma, S.B. & Rosenberg, N.J. (1983). Microclimate in the soybean canopy. *Agricultural Meteorology*, **28**, 321–37.

Bazzaz, F.A. (1979). The physiological ecology of plant succession. *Annual Review of Ecology and Systematics*, **10**, 351–71.

Beatley, J.C. (1969). Biomass of desert winter annual plant populations in southern Nevada. *Oikos*, **20**, 261–73.

Begg, J.E. & Jarvis, P.G. (1968). Photosynthesis in Townsville Lucerne (*Stylosanthes humilis* H.B.K.). *Agricultural Meteorology*, **5**, 91–109.

Begg, J.E. & Torssell, B.W.R. (1974). Diaphotonastic and parahelionastic leaf movements in *Stylosanthes humilis* H.B.K. (Townsville Stylo). In *Mechanisms of Regulation of Plant Growth*, eds. R.L. Bieleski, A.R. Ferguson & M.M. Cresswell, pp. 277–83. Bulletin 12. Wellington: The Royal Society of New Zealand.

Black, C.C. (1973). Photosynthetic carbon fixation in relation to net CO_2 uptake. *Annual Review of Plant Physiology*, **24**, 253–86.

Blad, B.L. & Baker, D.G. (1972). Orientation and distribution of leaves within soybean canopies. *Agronomy Journal*, **64**, 26–9.

Bonhomme, R., Varlet Grancher, C. & Artis, P. (1974). Utilisation de l'énergie solaire par une culture de *Vigna sinensis*. II. Assimilation nette et accroissement de matière sèche, influence du phototropisme sur la photosynthèse des première feuilles. *Annales Agronomiques*, **25**, 49–60.

Bradford, K.J. & Hsiao, T.C. (1982). Physiological responses to moderate water stress. In *Encyclopedia of Plant Physiology*, Volume 12B, eds. O.L. Lange, P.S. Nobel, C.B. Osmond & H. Ziegler, pp. 263–324. New York: Springer–Verlag.

Dubetz, S. (1969). An unusual photonastism induced by drought in *Phaseolus vulgaris*. *Canadian Journal of Botany*, **47**, 1640–1.

Ehleringer, J.R. & Forseth, I.N. (1980). Solar tracking by plants. *Science*, **210**, 1094–8.

Ehleringer, J.R. & Hammond, S.D. (1987). Solar tracking and photosynthesis in cotton leaves. *Agricultural and Forest Meteorology*, **39**, 25–35.

Ehleringer, J.R. & Mooney, H.A. (1983). Photosynthesis and productivity of desert and Mediterranean-climate plants. In *Encyclopedia of Plant Physiology*, Volume 12D, eds. O.L. Lange, P.S. Nobel, C.B. Osmond & H. Ziegler, pp. 205–31. New York: Springer–Verlag.

Fisher, F.J.F. & Fisher, P.M. (1983). Photosynthetic patterning: a mechanism for sun tracking. *Canadian Journal of Botany*, **61**, 2632–40.

Forseth, I.N. & Ehleringer, J.R. (1980). Solar tracking response to drought in a desert annual. *Oecologia*, 44, 159–63.
Forseth, I.N. & Ehleringer, J.R. (1982). Ecophysiology of two solar tracking desert winter annuals. II. Leaf movements, water relations and microclimate. *Oecologia*, 54, 41–9.
Forseth, I.N. & Ehleringer, J.R. (1983a). Ecophysiology of two solar tracking desert winter annuals. III. Gas exchange responses to light, CO_2, and VPD in relation to long-term drought. *Oecologia*, 57, 344–51.
Forseth, I.N. & Ehleringer, J.R. (1983b). Ecophysiology of two solar-tracking desert winter annuals. IV. Effects of leaf orientation on calculated daily carbon gain and water use efficiency. *Oecologia*, 58, 10–18.
Forseth, I.N. & Teramura, A.H. (1986). Kudzu leaf energy budget and calculated transpiration: the influence of leaflet orientation. *Ecology*, 67, 564–71.
Fukai, S. & Loomis, R.S. (1976). Leaf display and light environments in row-planted cotton communities. *Agricultural Meteorology*, 17, 353–79.
Gates, D.M. (1962). *Energy Exchange in the Biosphere*. New York: Harper & Row.
Gates, F.C. (1916). Xerofotic movement in leaves. *Botanical Gazette*, 61, 399–407.
Givnish, T.J. (1978). On the adaptive significance of compound leaves, with particular reference to tropical trees. In *Tropical Trees as Living Systems*, eds. P.B. Tomlinson & M.H. Zimmerman, pp. 351–80. Cambridge University Press.
Herbert, T.J. (1984). Axial rotation of *Erythrina herbacea* leaflets. *American Journal of Botany*, 71, 76–9.
Herbert, T.J. & Larsen, P.B. (1985). Leaf movement in *Calathea lutea* (Marantaceae). *Oecologia*, 67, 238–43.
Hsiao, T.C. (1973). Plant responses to water stress. *Annual Review of Plant Physiology*, 24, 519–70.
Kawashima, R. (1969a). Studies on the leaf orientation-adjusting movement in soybean plants. I. The leaf orientation-adjusting movement and light intensity on leaf surface. *Proceedings of the Japanese Crop Science Society*, 38, 718–29.
Kawashima, R. (1969b). Studies on the leaf orientation-adjusting movement in soybean plants. II. Fundamental pattern of the leaf orientation-adjusting movement and its significance for the dry matter production. *Proceedings of the Japanese Crop Science Society*, 38, 730–42.
Koller, D. (1981). Solar tracking (phototropism) in leaves of *Lavatera cretica* and *Malva parviflora*. *Carnegie Institute of Washington Yearbook*, 80, 72–5.
Lang, A.R.G. (1973). Leaf orientation of a cotton plant. *Agricultural Meteorology*, 11, 37–51.
Lang, A.R.G. & Begg, J.E. (1979). Movements of *Helianthus annuus* leaves and heads. *Journal of Applied Ecology*, 16, 299–305.
Ludlow, M.M. & Björkman, O. (1984). Paraheliotropic leaf movement in Siratro as a protective mechanism against drought-induced damage to primary photosynthetic reactions: damage by excessive light and heat. *Planta*, 161, 505–18.
McMillen, G.G. & McClendon, J.H. (1979). Leaf angle: An adapative feature of sun and shade leaves. *Botanical Gazette*, 140, 437–42.
Meyer, W.S. & Walker, S. (1981). Leaflet orientation in water-stressed soybeans. *Agronomy Journal*, 73, 1071–4.
Mooney, H.A. & Ehleringer, J.R. (1978). The carbon gain benefits of solar tracking in a desert annual. *Plant, Cell and Environment*, 1, 307–11.
Oosterhuis, D.M., Walker, S. & Eastham, J. (1985). Soybean leaflet movements as an indicator of crop water stress. *Crop Science*, 25, 1101–6.

142   J.R. EHLERINGER AND I.N. FORSETH

Rawson, H.M. (1979). Vertical wilting and photosynthesis, transpiration, and water use efficiency of sunflower leaves. *Australian Journal of Plant Physiology*, **6**, 109–20.

Satter, R.L. & Galston, A.W. (1981). Mechanisms of control of leaf movements. *Annual Review of Plant Physiology*, **32**, 83–110.

Schulze, E.–D. & Hall, A.E. (1982). Stomatal responses, water loss and CO_2 assimilation rates of plants in contrasting environments. In *Encyclopedia of Plant Physiology*, Volume 12B, eds. O.L. Lange, P.S. Nobel, C.B. Osmond & H. Ziegler, pp. 181–230. New York: Springer–Verlag.

Schwartz, A. & Koller, D. (1978). Phototropic response to vectorial light in leaves of *Lavatera cretica* L. *Plant Physiology*, **61**, 924–8.

Shackel, K.A. & Hall, A.E. (1979). Reversible leaflet movements in relation to drought adaptation of cowpeas, *Vigna unguiculata* (L.) Walp. *Australian Journal of Plant Physiology*, **6**, 265–76.

Shell, G.S.G. & Lang, A.R.G. (1976). Movements of sunflower leaves over a 24-h period. *Agricultural Meteorology*, **16**, 161–70.

Shell, G.S.G., Lang, A.R.G. & Sale, P.J.M. (1974). Quantitative measure of leaf orientation and heliotropic response in sunflower, bean, pepper, and cucumber. *Agricultural Meteorology*, **13**, 25–37.

Sheriff, D.W. & Ludlow, M.M. (1985). Diaheliotropic responses of leaves of *Macroptilium atropurpureum* cv. Siratro. *Australian Journal of Plant Physiology*, **12**, 151–71.

Stanhill, G. (1962). The effect of environmental factors on the growth of alfalfa in the field. *Netherland Journal of Agricultural Science*, **10**, 247–53.

Toft, N.L. & Pearcy, R.W. (1982). Gas exchange characteristics and temperature relations of two desert annuals: a comparison of a winter-active and a summer-active species. *Oecologia*, **55**, 170–7.

Travis, R.L. & Reed, R. (1983). The solar tracking pattern in a closed alfalfa canopy. *Crop Science*, **23**, 664–8.

Vogelmann, T.C. (1984). Site of light perception and motor cells in a sun-tracking lupine (*Lupinus succulentus*). *Physiologia Plantarum*, **62**, 335–40.

Vogelmann, T.C. & Björn, L.O. (1983). Response to directional light by leaves of a sun-tracking lupine (*Lupinus succulentus*). *Physiologia Plantarum*, **59**, 533–8.

Wainwright, C.M. (1977). Sun tracking and related leaf movements in a desert lupine *Lupinus arizonicus*. *American Journal of Botany*, **64**, 1032–4.

Werk, K.S., Ehleringer, J.R., Forseth, I.N. & Cook, C.S. (1983). Photosynthetic characteristics of Sonoran Desert winter annuals. *Oecologia*, **59**, 101–5.

Wien, H.C. & Wallace, D.H. (1973). Light-induced leaflet orientation in *Phaseolus vulgaris* L. *Crop Science*, **13**, 721–4.

Wofford, T.J. & Allen, F.L. (1982). Variation in leaflet orientation among soybean cultivars. *Crop Science*, **22**, 999–1004.

Yin, H.C. (1938). Diaphototropic movement of the leaves of *Malva neglecta*. *American Journal of Botany*, **25**, 1–6.

J.R. PORTER

8. Modules, models and meristems in plant architecture

Introduction: modules and architecture

The shape of the canopy influences many important aspects of the growth and development of plants and such effects are felt at many levels. Differences in canopy form may affect not only how much photosynthetically active radiation is intercepted by plants but may also regulate the spectral composition of radiation that filters to lower levels in the canopy and thus have photomorphogenetic consequences. The extent of shading both by and from close neighbours will also be affected by canopy shape, as will the degree of presentation to, or concealment from, consumers of nutritious foods such as fruits, leaves and buds. In a more agricultural context, canopy arrangement influences the extent to which disease spores or the droplets of a chemical designed to kill them (or prevent their development) can enter infectable zones.

The above ecological repertoire of plants is linked directly to their gross form and invites an obvious question concerning their evolution, namely: does the architectural 'type' of a plant have a rôle in the (Darwinian) fitness of an individual or, in other words, have certain whole plant forms been selected during evolution while others have been less successful?

This question forms the major theme of this chapter although Fisher (1984) has recently considered a similar topic. In addition, in order to better understand the mechanisms behind the magnificent variability in plant form that we see, some recent experimental data indicating the rôle of genomic changes in determining plant shape will be presented. This is followed by an example of the extent to which variability in canopy architecture can be induced by changes in the environment. Firstly, it is necessary to consider the basis from which differences in whole plant morphology arise.

Besides its 'functional' features, canopy shape has a taxonomic rôle in distinguishing one species from another. Reference to a common field guide for trees (e.g. Mitchell, 1974) shows that canopy outline may be as important when deciding an individual's specific identity as the conventional criteria of floral morphology and/or leaf shape. For example, the outline of an oak tree differs from that of a birch tree even though the general type of the component pieces (or modules, i.e. buds, leaves, shoots) is essentially the same in each case. Differences in whole-plant form are caused by the leaves and other plant parts being held in certain positions by branches of various lengths and at different angles. Also, and most importantly, the *number* of these modules, their rates of production and reduction, is a prime determinant of differences in form. In contrast, for most animals, changes in overall form during growth, and differences in form between mature animals arise not from any such changes to a branching, essentially repetitive framework, but from small accumulated modifications in the shape of particular organs or structures over time. Certainly, a component underlying the differences between two plant canopies may be that their leaves are not isomorphic between species or varieties. However, features such as the different number and positional geometry of the leaves and shoots are the over-riding cause of such diversity. Arber (1950) expressed this notion clearly: 'Amongst plants, form may be held to include something corresponding to behaviour in the zoological field... for most, though not all plants the only available form of *action* are either growth, or discarding of parts, both of which involve a change in the size and form of the organism.'

Morphic transformations in animals have been extensively investigated, most elegantly by D.W. Thompson (1917) in his seminal work *On Growth and Form*, and later by J.S. Huxley (1932). Such analytical insight as Thompson performed for morphology in whole animals has, unfortunately, little relevance for plants; a fact implicitly recognised by the sole botanical analysis in his book being the relationship between the mathematical Fibonacci series (1,1,2,3,5,8,...) and phyllotaxy. However, the study of whole plant form has been revived in recent years, after being a popular topic for botanical research in the eighteenth and nineteenth centuries, mostly via the Continental School (Mayr, 1982).

Models and architecture

An important step in this revival took place in the early 1970s when a fresh interest developed in the study of the form of plants. For previous work see the detailed reviews by White (1979, 1984). The most interesting features of this renaissance are (1) a re-discovery of the notion of modular construction and its importance in the generation of plant shape, and (2) an emphasis on understanding the mechanisms behind the dynamics of the production, orientation and turnover of plant parts (buds, flowers, roots) so that morphology can turn away from being associated with static taxonomical analysis to something with a more ecologically functional view (White, 1984).

The principal stimulus to this resurgence came with the publication, in French, of an essay by Hallé & Oldemann in 1970 on the architecture and the dynamics of growth of tropical trees, principally those in the forests of French Guiana (Hallé & Oldemann, 1970), later translated into English. This work and the fuller account which has followed (Hallé, Oldemann & Tomlinson, 1978) categorises the large-scale architectural diversity observed in the forest into a series of 23 architectural types or models (each reverently named after a distinguished botanist) on the basis of features of the gross branching system and not on conventional taxanomic criteria. Their models also include tree forms from temperate regions. Thus in a sense their work is an attempt to classify, but the bases for the taxonomy of their models are not conventional Linnean criteria (floral morphology, leaf shape, hairiness, etc.) and a single 'model' is often pan-specific and can even include members from different families of plants. Hallé & Oldemann also stressed the concept of 'reiteration' as an important principle in plant ontogenesis (Fig. 8.1). Reiteration is the addition to an initial tree of a new shoot system which conforms, in general, to the architectural model exhibited by the parent tree. However, this new shoot complex, although a scaled-down and faithfully reproduced version of the original model is not the product of a sexual process (cf. genet and ramet; Harper, 1977). Thus Hallé and his co-workers have both provided a formalised initial descriptive framework for a new look at whole-plant morphology and suggested simple first-order mechanisms by which plant forms generate themselves. Similarly, root systems are beginning to be classified via their architectural construction but it is felt (Barlow, 1986) that such a classification has not, as yet, reached the sophistication of the Hallé & Oldemann system.

Model definition

Most of the criteria used by Hallé, Oldemann and Tomlinson (1978) as the basis for describing their unique architectural models are dichotomous. They distinguish, among other characteristics (Table 8.1), shoots as being monopodial or

Fig. 8.1. Diagrammatic illustrations of reiteration in trees (*a*) reiteration on an orchard trunk; (*b*) stem suckers on an older tree (redrawn from Hallé, Oldemann & Tomlinson, 1978).

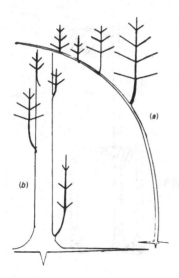

Table 8.1. *Terms referred to in the text, as defined by Hallé, Oldemann & Tomlinson (1978)*

monopodial (growth):	growth by the continued activity of a single meristem
sympodial (growth):	growth from successive lateral meristems
orthotropic (shoot):	a vertical axis resulting from a particular gravitational response of shoots
plagiotropic (shoot):	an oblique or horizontal axis
pleonanthic (shoot):	an axis which is not determinate by flowering i.e. flowers or inflorescences are laterally positioned
hapaxanthic (shoot):	an axis which is determinate by terminal flowering

Fig. 8.2. Schematic examples of an orthotropic (vertical): (*a*) monopodial shoot; (*b*) sympodial shoot; (*c*) hapaxanthic shoot; (*d*) pleonanthic shoot. Plagiotropic (horizontal) versions of these shoot types also exist.

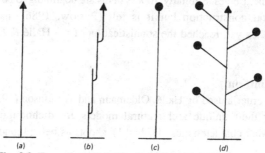

(a)   (b)   (c)   (d)

Fig. 8.3. Examples of McClure's model: (*a*) *Polygonum cuspidatum* Sieb. and Z.; (*b*) *Bambusa vulgaris* Schrader ex. Wendland (redrawn from Hallé, Oldemann & Tomlinson, 1978).

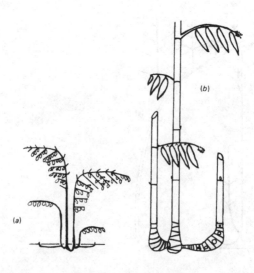

sympodial; orthotropic or plagiotropic; pleonanthic or hapaxanthic; basitonically or acrotonically branching and whether growth is rhythmic (i.e. with periods of dormancy) or continuous. Schematic representations of these terms are given in Fig. 8.2.

Figs. 8.3 and 8.4 illustrate two of Hallé and Oldemann's models. The first, McClure's model is shown, for example, in *Polygonum cuspidatum* Sieb. and Z. (Polygonaceae) (Fig. 8.3a), a common temperate weed from the Far East which grows up to 2 m in height and *Bambusa vulgaris* Schrader ex. Wendland (Graminae–Bambusoideae) (Fig. 8.3b) which is only known in cultivation. McClure's model is described as having a monopodial main shoot (MS), which may be orthotropic and hapaxanthic, and monopodial lateral branches (LB) which are plagiotropic and hapaxanthic. This model represents many members of the bamboo family (Bambusoideae) (McClure, 1966).

In Troll's model (Troll, 1937), all axes are sympodial, pleonanthic and plagiotropic. Examples are *Chrysophyllum cainito* L. (Sapotaceae, the 'star apple') (Fig. 8.4a) and *Anaxagorea acuminata* (Dun.) St. Hilaire (Annonaceae) (Fig. 8.4b), a small tree of the forest undergrowth which attains about 6 m in height and exemplifies the main model for the family Annonaceae.

Model extrapolation

Using these simple dichotomous criteria of shoots being monopodial/sympodial (growth by apical or lateral meristems); hapaxanthic/pleonanthic (lateral or terminal flowers); orthotropic/plagiotropic (vertical or horizontal axes) and distinguishing main shoots and branches in this binary scheme, it is possible to fit 18 of the 23 Hallé & Oldemann models into a two-way table (Table 8.2). The missing five models are excluded from such a format since they require additional criteria to distinguish them. There are two notable features in Table 8.2. Ignoring, for the most

Fig. 8.4. Examples of Troll's model: (*a*) *Chrysophyllum cainito* L.; (*b*) *Anaxagorea acuminata* (Dun.) St. Hilaire (redrawn from Hallé, Oldemann & Tomlinson, 1978).

Table 8.2. *Placement of* 18 *of the* 23 *Halle and Oldemann models within a dichotomous key for main shoots* (MS) *and branches* (B). X, *no model possible* (F) *fossil trees,* (*) *possible but unobserved models*

			MONOPODIAL MS				SYMPODIAL MS			
			HAPAXANTHIC MS		PLEONANTHIC MS		HAPAXANTHIC MS		PLEONANTHIC MS	
			ORTHO MS	PLAGIO MS	ORTHO MS	PLAGIO MS	ORTHO MS	PLAGIO MS	ORTHO MS	PLAGIO MS
SYMPODIAL B	PLEONANTHIC B	PLAGIO B	*	X	Auberville	X	Nozeran	X	Nozeran	Troll
		ORTHO B	*	X	*	X	Tomlinson	X	Tomlinson	X
	HAPAXANTHIC B	PLAGIO B	Petit	X	*	X	Prevost	X	Nozeran	X
		ORTHO B	Stone Fagerlind	X	*	X	Koriba (F) Tomlinson Leewenberg	X	Tomlinson	X
MONOPODIAL B	PLEONANTHIC B	PLAGIO B	Roux (F)	X	Schoute (F)	Schoute (F)	Nozeran	X	Champagnet	Champagnet
		ORTHO B	Roux (F)	X	Rauh (F) Schoute Attims	Schoute (F)	*	X	Mangenot	Mangenot
	HAPAXANTHIC B	PLAGIO B	McClure	X	Massart (F)	X	Nozeran	X	Nozeran	X
		ORTHO B	*	X	*	X	*	X	*	X

Fig. 8.5. Examples of fossil trees: (*a*) *Lepidophloios sp;* (*b*) *Archaeopteris macilenta* (redrawn from Hallé, Oldemann & Tomlinson, 1978).

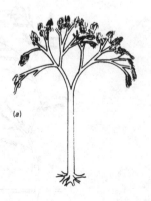

(a)

(b)

part, those categories which have plagiotropic MS (although examples are seen of models in which the MS can be plagiotropic but change to orthotropic or *vice versa*, i.e. Troll's model and Mangenot's model) it can be seen that fossil plants (F) mostly occupy cells describing Roux's model and Schoute's model with some members in Koriba's model. Examples are *Lepidophloios* sp. Lepidodendrales – Lepidodendraceae (Schoute's model), Sigillariaceae (Schoute's model) (Fig. 8.5*a*) and *Archaeopteris macilenta*, Archaeopteridales (Roux's model) (Fig. 8.5*b*). These plants of varying stature are characterised by a growth-form that has a monopodial MS with monopodial LB, both hapaxanthic and pleonanthic MS but predominantly pleonanthic LB and ortho- and plagiotropic LB. There is a pronounced lack of fossil examples in models where sympodial branching is observed and where both pleonanthic and hapaxanthic flowering occur.

The clumping of fossil trees amongst the Hallé & Oldemann models suggests that some plant forms may have paid an evolutionary penalty for their mode of whole plant development.

In addition, Table 8.2 is not complete. There are gaps in the table where nature has seemed not to place certain combinations of morphological pattern.

If one draws these forms, according to the simple scheme described in Fig. 8.2, the strangeness of these plants is evident (Fig. 8.6). The first (Fig. 8.6*a*) with a monopodial, pleonanthic and orthotropic MS and sympodial, hapaxanthic and plagiotropic LB seems to be over-laden with floral parts at the expense of any possibilities of vegetative growth. Fig. 8.6(*b*) shows a very strange looking individual that does, however, bear some resemblance to Nozeran's model. Similar analyses

Fig. 8.6. Hypothetical tree-like forms from the Hallé & Oldemann system: (*a*) monopodial, pleonanthic, orthotropic MS with sympodial, hapaxanthic, plagiotropic LB; (*b*) sympodial, pleonanthic, orthotropic MS with monopodial, hapaxanthic and orthotropic LB.

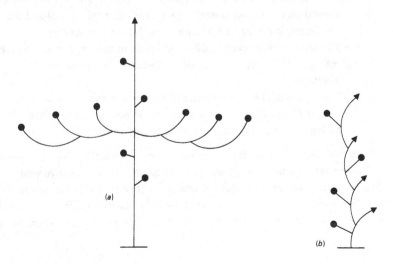

have been performed by Niklas (1986) who has described a finite suite of three-dimensional plant forms based on all possible permutations of the extreme values of branch angle, branching probability and angular rotation with respect to parent branches, that have been observed. All observed plant forms exist within such a three-dimensional space and thus non-observed plant forms can also be generated.

Adaptation of different models

Nor surprisingly the adaptive significance of plant form has interested Hallé's students (and others, notably Horn, 1971; and Fisher, 1984). One of them, Fournier (1979), examined arguments for and against the idea that the different architectural patterns, defined by Hallé & Oldemann's models, have adaptive consequences for the plant. I will only present the conclusions to her arguments.

Support for the idea that plant shape is adaptive comes from the observations that:

(1) There are species that display different models for different environments e.g. *Laurus nobilis* L. (Lauraceae) displays Rauh's model in brightly lit conditions but Massart's model in the shade (Ramarosan–Ramparany, 1978 cited in Fournier, 1979). Other specific switches between different models have also been noted in response to changes in irradiation.

(2) There is an increased frequency of certain models exhibiting a low branching habit amongst high tropical mountain plants – suggesting protection against frost or drought.

Fournier also offers the opposing argument with the following observations, that tree form is not adaptive;

(1) All models exist in lowland tropical rain forests (where Hallé & Oldemann started their work) suggesting that a single ecological region has *not* favoured some models at the expense of others. However, as a habitat, the tropical rain forest possesses many niches (Richards, 1952) and thus there may be ample ecological space for examples of all the models.

(2) The same models exist at different levels in the forest such that vertical gradients of radiation, predation and nutrition seem not to influence their distribution.

(3) The same model exists in different life forms from trees 40 m high to small ruderal herbs. Clearly, such extremes of biotype do not share the same ecology and thus their 'model type' is irrelevant.

From these arguments and from the evidence presented earlier about fossil trees and the 'unseen' plant forms, it can be seen that the question of the selective value of plant form, in the narrow sense in which it was asked (Fisher, 1984) is still unanswered. Form, however, does have a role to play in evolution. Darwin (1859) in *On the Origin of Species* stressed its importance succinctly and powerfully when he stated

'[Morphology] is the most interesting department of natural history and may be said to be its very soul', but it may be other features of form such as scale and/or size or reproductive biology (Maynard Smith, 1978) or any capacity for regeneration ('reiteration' *as per* Hallé & Oldemann) that may be more important than the taxonomic architectural types recognised by Hallé and his co-workers. It may even be that a 'successful' canopy is one that is simply of a different shape to that of close neighbours. Such a strategy of maximising canopy heterogeneity, the making of a speciality out of non-specialisation, is perhaps the key to canopy shape. However, this hypothesis if it were true, would lead to the theoretical conclusion that it is merely the appearance of fresh architectural heterogeneity and not natural selection for particular architectural features that moulds the variety of whole plant shapes that we see.

Meristems and architecture

Hallé & Oldemann's idea of large-scale morphological *leitmotivs* for, in their words, 'plant-making' is a stimulating one but suffers, as with all taxonomy, of being necessary as the starting point of a morphological study but limiting as its end point. There is a need to delve deeper into the controlling mechanisms of plant form.

The second question posed at the beginning of this chapter concerned the causal basis of the differences in form between one plant and another. There are many suggested 'causes' for plant form, such as the view that plants are units of constructional engineering where the shapes of their branches, their elasticity and resistance to strain, are constrained by well-known mechanical principles (McMahon & Kronauer, 1976). Closely allied to this view is the theoretical model of Shinozaki *et al.* (1964*a,b*) who interpreted plants as being constructed of vertical and horizontal fascicular pipes with each pipe responsible for linking a particular piece of root system to a portion of the leaf canopy. The diameter of a pipe is a function of the weight of canopy to be supported by it and more pipes are added as the plant ages. Also there are those who contend that plant form is an almost passive response to evolutionary pressures exerted by herbivory, predation and disease (Harper, 1984) and that, by being modular organisms, plants are uniquely fitted to cope with such onslaughts. I take the view that plant form is initially best understood in terms of plant modularity and that we may look upon whole plants as fragmentary individuals in which we describe the birth, ageing and death, of separate plant meristems and then *resynthesise*, with simulation models for example, representations of the whole organism not only as it is now but how it might become (e.g. Bell, 1984). There is no real equivalent to this possibility in animals given their overall life form. What we mean by the 'habit' or 'shape' of a plant occurs as the result of interaction between two parameters of modular growth, namely the propensity to produce other modules, which occurs in time and has a measurable rate, and the positioning of these modules – which introduces a spatial dimension to the process. Plant form is a direct consequence of these dynamics and, by using them to explain form, a mechanistic basis for the shape of whole plants becomes possible. The simplest scheme that incorporates the above is shown in Fig. 8.7 (Porter, 1983a). For a plant containing both vegetative and flowering meristems, over discrete periods of time (Δt) a vegetative meristem can remain as such (i.e. stay dormant) or can grow to produce copies of itself. Such replication is not unbounded (Fresco, 1973; Porter, 1983b; White, unpublished) for presumably the same reason that competition for resources curtails the exponential growth of populations of individuals.

If some of the meristems on a shoot are flowering meristems then over Δt they can remain as such, or they can die, or they can develop into mature floral units. Flowers either live on into the next unit of time, or they die, or they set seed. Thus the system can be summarised as a series of transitional probabilities for the components as they

Fig. 8.7. The positive transitions in a plant that produces both vegetative and flowering meristems. Dashed lines indicate how new meristems are added to the total. VM, vegetative meristem; VBR, vegetative branch; FM, flowering meristem; DFM; dead flowering meristem; FBR, flower; DFBR; dead flower; t, time (arbitrary units); Δt, discrete increment of time (after Porter, 1983b).

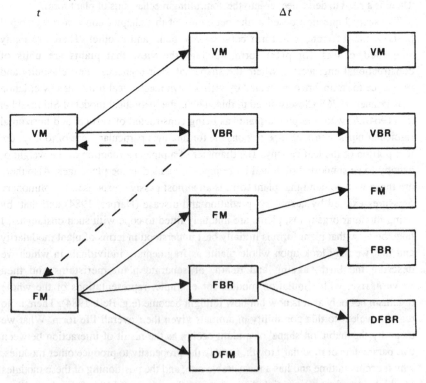

move from one state to another. We interpret these transitions and the accompanying change in number of parts as whole plant development.

Studies based on the preceding analysis have been carried out (Porter, 1983*b*). Notable is the recent work of van Groenendael (1985) on *Plantago lanceolata* L. who simulated its growth from rosette to flowering plant by four simple instructions applied iteratively to the metameric units of leaf plus internode plus axillary bud. The basic model is shown in Fig. 8.8. The leftmost cycle of operations describes the iterative operations necessary for generating the normal vegetative rosette of a plantain. Step 1 is to form an internode and suppress its elongation; step 2 is to form a node and a normally shaped leaf; step 3 is to form an axillary meristem; and step 4 is to rotate the metamer by a few degrees. Activation of axillary meristems will lead to side rosettes, via the 'building rules' in the other circle. To model the development of flowers requires further instructions to the metamers, as shown to the right in Fig. 8.8.

As a genus, *Plantago* is well known for possessing a variety of teratologies – developmentally based malformations – within its species. Van Groenendael extended his application of the above rules of metameric development to describe further the development of these abnormalities. He found two types of rule breakage that

Fig. 8.8. Sequence of events for the production of a single flowering rosette of a plantain via metameric units. Numbers refer to the construction steps (see text).

Key: *, germination; ──────, developmental path in first order meristem; ══,
developmental path in second-order meristem; ═══ developmental path in third
order meristem; (▶), compulsory step; (▷), optional step (after van Groenendael,
1985).

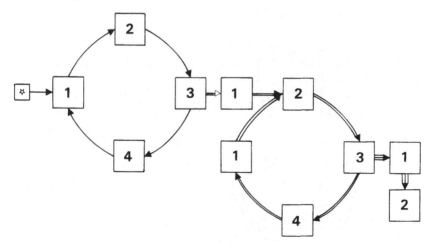

produced such developmentally aberrant individuals (Fig. 8.9). Errors occur via 'wrong steps'; that is, the sequence of execution of steps is correct but a step is executed wrongly or missed out. For example the plants indicated in Fig. 8.9(*a, b* and *c*) went wrong at step 1; those at Fig. 8.9(*d, e* and *f*) failed at step 2 – and so on. Other developmental errors occur when sequences of building rules are executed correctly but at the wrong time. For example, in non-teratological individuals, flowering spikes come from second-order meristems. For plants to have a central flowering spike produced by first-order meristem means that the correct 'flower' instructions have been followed but at the wrong time in the life of the plant. Combinations of these two types of error combine to produce delightfully exotic individuals, Fig. 8.9(*j, k* and *i*)! Van Groenendael's work illustrates how by focusing on the units of construction and the rules and mistakes that can occur in their implementation a degree of mechanistic understanding of the variation in plant form can be established. Linking such a dynamic approach to plant form to the approach of Hallé & Oldemann is both evident and clearly necessary as is the possibility of identifying the genes responsible for such malformations (van Groenendael, personal communication).

Fig. 8.9. Teratological individuals in *Plantago*: (*a–f*) results of wrong steps in model (Fig. 8.8 and text); (*g–i*) results of wrong order in model execution; (*j–l*) combination of wrong steps and wrong orders (after van Groenendael, 1985).

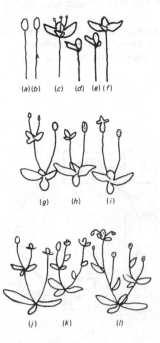

Environmental architecture

The above examples show how heritable malfunctions can alter the fundamental type of a model. However, there are conditions in which the expression of a particular model is, although unchanged in a qualitative sense, greatly modified by the environmental conditions in which the plants grow (Hallé, 1976).

An important question is how extreme can differences in the quantitative expression of a particular model be whilst still recognisable as the same model. The answer seems to be for McClure's model, that wide variation is allowed in its quantitative, environmentally controlled, expression and that the same overall leaf canopy can arise via very different shoot hierarchies, each faithfully exhibiting a particular model-type.

Winter wheat (*Triticum aestivum*), in common with many members of the Gramineae, belongs to McClure's model since it has monopodial, terminally flowering shoots that are orthotropic for the main shoots and initially plagiotropic for lateral shoots. In this respect it is the same as the bamboo plant in Fig. 8.2.

When two crops of winter wheat were sown at the same seed density but on dates separated by just over a month (Willington & Biscoe, 1982) their green area indices (GAI), a measure of the coverage by the leaves of the soil surface, were very similar (Fig. 8.10a), but the pattern of tillering was markedly different (Fig. 8.10b). For the earlier sown crop a sizeable component of the shoot population was made up of second- and higher-order tillers which were absent for the later sown crop. In the latter situation tiller 2 made the second largest contribution, after the main shoots, to the total shoot population. Interestingly, the initial, maximum and final number of shoots for both situations were very similar but the composition of the population was very different.

When Willington & Biscoe (1984) varied both the amount and timing of nitrogen fertiliser applied to the crop (for a single sowing date and sowing rate), very different canopies were produced. The difference in GAI between the treatments in which 250 kg ha^{-1} of nitrogen was applied as a single dressing or in three portions is seen in Fig. 8.11a. However, the corresponding patterns of tillering for the two treatments were extremely similar (Fig. 8.11b). In one environment the same tiller hierarchy produced a different total leaf cover, whereas other sets of growing conditions resulted in the same leaf canopy, even though there were vastly different tiller populations. Thus the extremes of expression of a model seem to be very wide, although it is rare to get such a precise illustration of what might be called 'architectural compensation' between the leafy and the leaf supporting components of a canopy. What such observations show is the wide plasticity allowable within one Hallé & Oldemann model, and that to understand the mechanisms by which canopies develop requires that we recognise that overall descriptors such as GAI, relative growth rate and total shoot number may obscure as much as they reveal about canopy development. They are outcomes of processes that have as their basis the birth, death and ageing of shoot and meristems.

Obtaining data for studying such systems usually requires a mixture of simple technical support and an ability to count to large numbers (e.g. $>10^4$) – a suitably frugal combination of budgetary requirements for these austere times. It is by focusing on plants as assemblages of metameric units that the complexity of their canopies will begin to be understood. This will lead to a union of the descriptive work of Hallé and his colleagues with a yet more mechanism-based understanding of the development of whole plant form.

Fig. 8.10(*a*) GAI for two crops sown on two sowing dates (S1, (●), S2, (■) at Brooms Barn Experimental Statio, Suffolk; (*b*) tillering patterns of the crops. MS, main shoots; T1, tiller 1; T2, tiller 2; T3, tiller 3; RT, other shoots (after Willington & Biscoe, 1982).

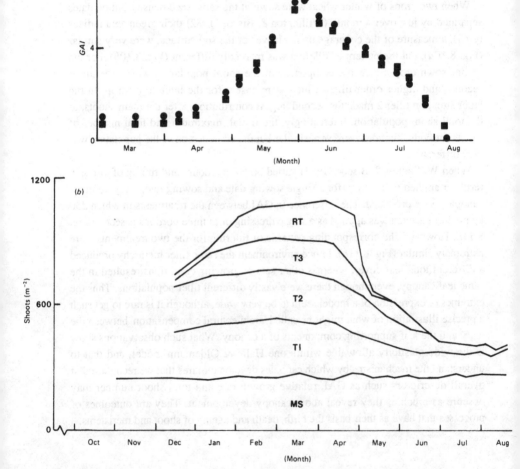

Fig. 8.11(*a*) *GAI* for two crops from February until final harvest where 280 kg ha^{-1} of nitrogen was applied single (●) or in three lots (■); *(b)* the stem classes are MS, T1, T2, RT respectively, as in Fig. 8.10.

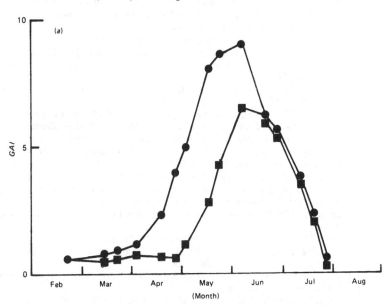

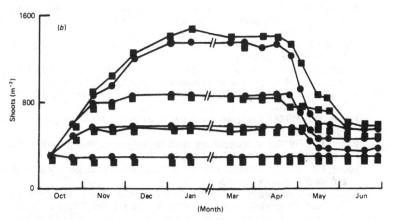

References

Arber, A. (1950). *The Natural Philosophy of Plant Form*. London: Cambridge University Press.

Barlow, P.W. (1986). Adventitious roots of whole plants: their forms, functions and evolution. In *New Root Formation in Plants and Cuttings*, ed. M.B. Jackson, pp. 67–110. Dordrecht, Boston, Lancaster: Martinus Nijhoff Publishers.

Bell, A.D. (1984). Dynamic morphology: a contribution to plant population ecology. In *Perspectives on Plant Population Ecology*, eds. R. Dirzo & J. Sarukhan, pp. 48–65. Sunderland, Mass.: Sinauer Associates Inc.

Darwin, C.R. (1859). *The Origin of Species by Means of Natural Selection; or the Preservation of Favoured Races in the Struggle for Life.* London: J. Murray.

Fisher, J.B. (1984). Tree architecture: relationships between structure and function. In *Contemporary problems in plant anatomy*, eds. R.A. White & W.C. Dickeson, pp. 541–89. Orlando, Florida: Academic Press.

Fournier, A. (1979). *Is Architectural Radiation Adaptive?* Diplome d'études approfondies d'écologie générale et appliqué. Montpellier, France: Université des Sciences et Techniques du Languedoc.

Fresco, L.F.M. (1973). A model for plant growth. Estimation of the logistic function. *Acta Botanica Neerlandica*, **22**, 486–9.

Hallé, F. (1976). Architectural variation at the specific level in tropical trees. In *Tropical trees as living systems*, eds. P.B. Tomlinson & M.H. Zimmerman, pp. 209–21. London, New York, Melbourne: Cambridge University Press.

Hallé, F. & Oldemann, R.A.A. (1970). *Essai sur l'Architecture et la Dynamique de Croissance des Arbres Tropicaux*. Paris: Masson. English translation by B.C. Stone, Penerbit University of Malaya, Kuala Lumpur, (1975).

Hallé, F., Oldemann, R.A.A. & Tomlinson, P.B. (1978). *Tropical Trees and Forests*. Berlin, Heidelberg, New York: Springer Verlag.

Harper, J.L. (1977). *Population Biology of Plants*. London, New York: Academic Press.

Harper, J.L. (1984). Introductory comments. In *Perspectives on Plant Population Ecology*, eds. R. Dirzo & J. Sarukhan, pp.xv–xviii. Sunderland, Massachusetts: Sinauer Associates Inc.

Horn, H.S. (1971). *The Adaptive Geometry of Trees*. Princeton, New Jersey: Princeton University Press.

Huxley, J.S. (1932). *Problems of Relative Growth*. London: Methuen.

Maynard Smith, J. (1978). *The Evolution of Sex*. Cambridge University Press.

Mayr, E. (1982). *The Growth of Biological Thought*. Cambridge, Mass.: Harvard University Press.

McClure, F.A. (1966). *The Bamboos, the Fresh Perspective*. Cambridge Mass.: Harvard University Press.

McMahon, T.A. & Kronauer, R.F. (1976). Tree structures: deducing the principle of mechanical design. *Journal of Theoretical Biology*, **59**, 443–66.

Mitchell, A. (1974). *A Field Guide to the Trees of Britain and Northern Europe.* London: Collins.

Niklas, K.J. (1986). Computer-simulated plant evolution. *Scientific American*, **254**, 68–75.

Porter, J.R. (1983a). A modular approach to analysis of plant growth. I. Theory and principles. *New Phytologist*, **94**, 182–90.

Porter, J.R. (1983b). A modular approach to analysis of plant growth. II. Methods and results. *New Phytologist*, **94**, 191–200.

Ramaroson–Ramparany, L. (1978). *Contribution a l'Etude Architecturale de Quelques Espèces Ligneuses de Regions Tempérées*. Thèse de spécialité. Montpellier. France: Université des Sciences et Techniques du Languedoc.

Richards, P.W. (1952). *The Tropical Rain Forest*. Cambridge University Press.

Shinozaki, K., Yoda, K., Hozumi, K. & Kira, T. (1964a). A quantitative analysis of plant form – the pipe model theory. I. Basic analyses. *Japanese Journal of Ecology*, **14**, 97–105.

Shinozaki, K., Yoda, K., Hozumi, K. & Kira, T. (1964b). A quantitative analysis of plant form – the pipe model theory. II Further evidence of the theory and its application in forest ecology. *Japanese Journal of Ecology*, B14, 133–9.

Thompson, D.W. (1917). *On Growth and Form*. Cambridge University Press.

Troll, W. (1937). *Vergleichende Morphologie der Hoheren Pflanzen,* Band 1, teil 1. Berlin: Borntraeger.

van Groenendael, J.M. (1985). Teratology and metameric plant construction. *New Phytologist,* **99**, 171–8.

White, J. (1979). The plant as a metapopulation. *Annual Review of Ecology and Systematics,* **10**, 109–45.

White, J. (1984). Plant metamerism. In *Perspectives on Plant Population Ecology,* eds. R. Dirzo & J. Sarukhan, pp. 15–47. Sunderland, Mass.: Sinauer Associates Inc.

Willington, V.B.A. & Biscoe, P.V. (1982). *Growth and Development of Winter Wheat.* Report of the ICI Agricultural Division Financed Research Programme, Brooms Barn Experimental Station.

Willington, V.B.A. & Biscoe, P.V. (1984). *Growth and Development of Winter Wheat.* Report of the ICI Agricultural Division Financed Research Programme, Brooms Barn Experimental Station.

J.M. NORMAN

9. Synthesis of canopy processes

Introduction

A synthesis of canopy processes can be accomplished at various levels of detail. If historical data are available, then a statistical analysis of that data may provide a kind of synthesis; however, in this case the synthesis is implicit in the statistical tool used, yielding limited insight to us. Alternatively, a mechanistic approach can be used and each relevant process described by appropriate, state-of-the-art, quantitative relations with explicit integration (or synthesis) to achieve an 'integrated whole'. Clearly, statistical and mechanistic approaches represent extremes of a continuum where all intermediate states are possible. Thus a clear statement of objectives, guiding rules for pursuing these objectives, definition of the system, and evaluation criteria are prerequisites for beginning an orderly synthesis of canopy processes.

This chapter represents an attempt at an orderly synthesis of canopy processes with a reasonably mechanistic approach. The plant-environment model entitled Cupid (Norman & Campbell, 1983; Norman, 1979; Norman, 1982) is used as an example.

Rules for constructing a model

A system of rules for pursuing a synthesis of processes can aid one in resisting the temptation to 'over sell' and thus avoid having either to resort to short-term expediency when failure is in sight, or to justify the means deceptively with an end result that was essentially known before the modelling was begun.

The following rules are essential for constructing a mechanistic model if it is to be widely applicable. It must:

(1) have an objective;
(2) parameterise mechanisms at spatial and temporal scales smaller than the scale of prediction;
(3) obtain boundary conditions, such as climatic data, and plant physiological response functions from disciplinary studies, so that they are not unique to the model;
(4) test the model against independent measurements made in the field and eliminate any 'disposable' parameters that can be adjusted to improve agreement;

(5) adopt a structured approach to the definition of the system being modelled through the use of a structured programming language and structured output;

(6) include all component processes in a single program code for true integration; this program may consist of many modules or sub-programs;

(7) use state-of-the-art concepts in each impinging discipline at the appropriate level of detail and avoid 'quick fixes' when a knowledge gap exists;

(8) where possible seek mechanisms which are common to a wide range of plants experiencing similar conditions.

An appropriate objective has short- and long-term aspects. Cupid has three objectives: (a) to integrate knowledge across disciplines to learn new things about plant-environment relations; (b) to obtain detailed quantitative descriptions of soil-plant-atmosphere interactions to add confidence to simpler operational models; (c) to define the canopy environment in sufficient detail that environmental characteristics, which may be too difficult to measure routinely but which have significant impact on physiological processes of interest, can be estimated from a few simple measurements. The first two objectives are self-evident, but the third contains the implication that one might use routine measurements and a mechanistic model to aid in physiological field studies, much as one might use a growth chamber in laboratory studies.

The parameterisation level of a model is the smallest explicit spatial or temporal scale used. It represents the integration step interval and the result of the integral is appropriate at the prediction level. Although empirical relations are used at the parameterisation level, the parameters in these relationships may be obtained from smaller-scale mechanistic models or by direct measurement. For example, in Cupid the parameterisation level is at the leaf scale (centimetre spatial scale) and the 15 min–1 h time scale, whereas the prediction level is at the field scale and seasonal time scale. Processes such as photosynthesis and stomatal functioning must be parameterised either by direct measurement in the field with gas exchange systems or by models such as those of Farquhar & von Cammerer (1982) or Farquhar & Wong (1984). The model of Farquhar & von Cammerer (1982) was used by Norman (1986) with a simpler coupling to stomatal functioning than that of Farquhar & Wong (1984).

Environmental boundary conditions which are relatively easy to obtain are weather conditions above the canopy (wind, radiation, air temperature and relative humidity and precipitation) and conditions below the root zone (soil water content and temperature). Conditions at the soil surface are inappropriate inputs because they are too difficult to obtain.

All parameter values in the model must be obtained independently of any comparison between model prediction and observation; that is, no parameters should be adjusted to make model predictions agree better with prediction-level measurements. Thus in Cupid if canopy-level photosynthesis predictions do not agree with canopy-level measurements, the leaf characteristics are not adjusted; the model itself must be questioned or new leaf data obtained. This is a demanding rule and its application depends on the integrity of the scientist. Often when a model is referred to as 'calibrated', this means that parameters were adjusted so that predictions matched observations; a significant weakness in any mechanistic model.

Mathematical relationships in the model should be of the same form as those used in the discipline from which they originated for the greatest generality and widest understanding and acceptance. Mathematical formulations unique to the model can become a refuge for the modeller. This rule is also demanding because linkages between disciplines may be made less difficult and experts in various disciplines can understand what the modeller is doing and thus become more vocal in their criticism of shortcomings.

The greatest weakness in implementing a mechanistic model is the jumble of program code that can result from the rush to achieve an output. A structured approach is the only answer. Care in the design with as many, if not more, comments as code, will save time in the long run, and the use of structured languages such as PASCAL will encourage good program design. However, there are no shortcuts to good programming.

Definition of a model system

The definition of a model system emerges from a tension between two opposing elements: an attempt to be faithful to the objectives, and the practicalities of trying to do this. A model is a simplification of reality, and as such the process of simplification involves informed judgment.

The spatial and temporal boundaries of the model must be defined, or one may become trapped in the ever-downward spiraling approach of the reductionist and never accomplish integration. Cupid has a smallest scale of a few centimetres and a largest scale of a few hundred metres in the horizontal. In the vertical the lower boundary is the bottom of the root zone and the upper boundary is from a few metres to a few tens of metres above the canopy in the atmospheric surface layer. The shortest time scale is about 15 min–1 hr and the longest is about a growing season (months).

These boundaries derive from certain conceptual limitations related to existing knowledge, as well as to human and financial resources, and these conceptual limitations necessitate various assumptions. For example, in Cupid all transfer processes are assumed to be essentially one-dimensional with vertical fluxes and gradients being most important.

The objectives of Cupid focus on plant-environment relations; thus this model emphasises mass, momentum and energy exchanges between a plant and its environment of soil and atmosphere. The mathematical description of energy exchanges begins at the organ level. Since the organ that most influences canopy exchange processes is usually the leaf, we will discuss the energy balance (Fig. 9.1) of leaves in different layers of the canopy and inclined at various angles to the direction of the sun (Norman, 1979). Solving the leaf energy budget, which involves radiative, sensible and latent heat exchanges, requires a knowledge of the radiative, temperature, humdity and wind environments adjacent to the leaf. Environmental conditions adjacent to leaves are obtained from one-dimensional radiation (deWit, 1965) and turbulent transfer equations (Goudriaan, 1977) which in turn need the collective results of the leaf energy budgets for all the leaves. Therefore leaf energy budgets and vertical profile equations must be solved simultaneously (Fig. 9.2). The following one-dimensional equation is used to describe the vertical transfer of any quantity:

$$\frac{\partial A}{\partial t} = \frac{\partial}{\partial z}(k\frac{\partial E}{\partial z}) + Q \; , \tag{1}$$

where k is a vertical transfer coefficient relating the concentration or potential gradient of the quantity E to a flux of that same quantity, z is height, t is time, Q is a source or sink of this same quantity per unit volume, and the term on the left of the equality is the change in the quantity E per unit volume per unit time. Eqn. (1) can be used to describe momentum, heat, water or CO_2 fluxes in the canopy, atmosphere or soil. The source or sink, Q, may be determined from leaf processes in the canopy, root uptake, phase changes or input of rain to the soil.

One conceptual limitation in considering the leaf level is that leaves are assumed to

Fig. 9.1. The components of the leaf energy budget included in the model Cupid (Norman, 1982).

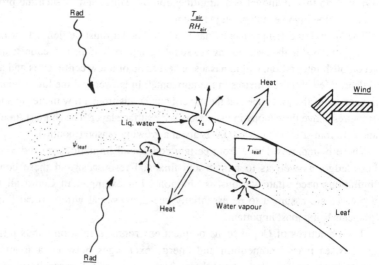

Leaf Energy Budget

be fixed and flat. A partially lit, flapping leaf in the wakes of other leaves has an extremely complex boundary layer which has been described but never quantified (Grace, Ford & Jarvis, 1981).

The description of a complex system often depends on identifying the important feedback loops between processes and preserving the conservation principles in

Fig. 9.2. The vertical exchange of energy, mass and momentum.

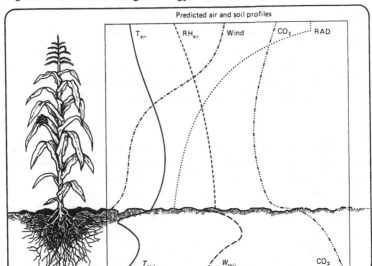

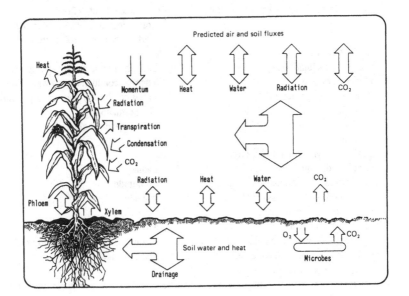

quantifying these feedbacks. Transpiration is an example of a process that can be controlled by feedback. The opening of stomata tends to increase the rate of transpiration, but such an increase in the rate would lower leaf temperature and raise canopy humidity, which tends to reduce the rate of transpiration. The net result of this negative feedback is that a doubling of stomatal conductance may only increase transpiration by 20%. This illustrates that the response of a system may depend more on the nature of the feedbacks than on the intrinsic characteristics of the individual components.

The soil and plant parameters essential to a model such as Cupid must be determined independently if we are to follow the rules set forth earlier. Table 9.1 contains a list of inputs required by Cupid. If procedures or instruments for obtaining the inputs are not available, then the rules presented dictate that such methods must be found before proceeding. Consider leaf area index (LAI), this can be a troublesome

Table 9.1. *Typical inputs for detailed plant–environment models such as Cupid*

Atmosphere (hourly)
Solar radiation
Air temperature
Air vapour pressure
Wind speed
Precipitation
amount
intensity
Rainfall or irrigation water temperature

Plant
Leaf area index
Mean leaf inclination angle
Plant height
Row spacing, plant spacing
Leaf size
Leaf spectral properties (PAR, solar, thermal)
Leaf photosynthetic rates (vs light, temp., vpd, water stress)
Leaf stomatal conductance (vs light, temp., vpd, water stress)
Leaf cuticular conductance
Plant hydraulic resistance
Precipitation interception characteristics (stem flow, wetted foliage area fraction, canopy storage amount)

Leaf surface characteristics affecting wettability of leaves
Root length density distribution and rooting depth

Soil
Lower boundary condition for soil temperature and soil water content at bottom of root zone (about 1–2m)
Soil reflectance (PAR, solar, thermal)
Vertical profile of soil texture and bulk density
Soil water content vs water potential
Thermal conductivity vs water potential
Water conductivity vs water potential
Gas permeability
Soil surface roughness

Other
Latitude
Longitude
Altitude
Date, time
Measurement heights and depths
Planting date or date of bud break
Site slope and aspect

input to obtain but it is essential and modelling without it is highly questionable. The author has worked for years to develop and test methods for obtaining LAI and mean leaf angle by rapid indirect methods. Thus we now can measure LAI and mean leaf angle in canopies varying in depth from 50 mm grass swards to 50 m forests relatively easily (Norman & Campbell, 1988). Therefore this variable should be obtainable for virtually all canopies in the future. Recent developments in portable, field gas exchange systems now make it possible to obtain the necessary stomatal and photosynthetic response functions of leaves needed for a model like Cupid.

Programming strategy

The program code of a complex model represents the greatest obstacle to dissemination of the ideas embodied in that model. Numerical analysis on the computer is a very powerful tool for modelling complex, natural systems, but the programs can be difficult, sometimes even for the originator, to understand. As such the originator often must be relied on to maintain the program code. Since the profession has no direct method for rewarding the originators of programs, other than by peer-review of published work, there is little incentive to maintain a program once the papers have been accepted by a journal. A very desirable alternative for dealing with documentation of complex models is that used by researchers in Wageningen, Netherlands; a PUDOC book describing the model has been published (Goudriaan, 1977).

The usual way of avoiding the service role of maintaining a program for others is to minimise documentation. Thus inadequate documentation is 'normal' for canopy models, including Cupid, because academics normally are penalised, albeit indirectly, for such service activity. The use of program structures, such as those in PASCAL, along with many comments within the code, allow the originator to work easily with very large programs and little documentation. Furthermore, such a well-written program, combined with definitions of input and output variables, constitutes a minimal level of documentation that will ensure that only colleagues seriously interested in the program will make demands on the originator.

Complex models are effective integrators when modellers cooperate closely with colleagues in complementary disciplines through joint research projects. Further, the modellers should be involved in measurement activities to maintain a balanced perspective.

Before embarking on a modelling program, some thought needs to be given to the following questions:

(1) How will continuity be maintained in the program code generated over years?

(2) How and at what level will documentation be obtained?

(3) How will this integrated knowledge be communicated to the research community; journal articles, book chapters, reports, books, etc.?

(4)   What programming language will be used?

An appropriate code can be generated by three means:

(1)   The principal scientist writes his own code. This has the advantage that he has a close working knowledge of the program so he can make modifications quickly and some continuity is maintained. It has the disadvantage, however, that the program is likely to be poorly transferable and difficult for anyone else to use.

(2)   A professional programmer is hired to generate 'more efficient' programs and thus free scientists to 'do what they do best', which presumably is not programming. A competent programmer will demand a clear specification of the overall objective of the model and its component sub-processes. This is an important step in any model and one which is often overlooked. If you cannot communicate your requirements to a programmer, you probably are not communicating your work effectively to anyone else. However, a word of caution is in order concerning programmers. Many programmers are not numerical analysts and proficiency with numerical methods is the key to developing mechanistic plant–environment models. Furthermore, how does a principal scientist who is not a programmer determine that he is working with a competent programmer at a sufficiently early stage?

(3)   Within academia, graduate students may be responsible for the code generation. This may be good experience for the students, but can be disaster for a program because of the short residence time of students along with the marginally acceptable code and documentation resulting from the learning process.

The code for the model Cupid is generated under the first of the above options.

Many programming languages are available but FORTRAN remains the most universal. Cupid is written in FORTRAN but with an attempt to use program structures inherent in the PASCAL language. Of course, FORTRAN does not require such structures so considerable discipline is required to keep the program from becoming an indecipherable quagmire.

Tests of models

A model is truly tested only when predicted results are compared with independent measurements. The word 'independent' is used here to mean measurements which have never been used in the formulation of the model. For example, if canopy transpiration is to be predicted then *no* measurements of canopy transpiration, even from other sites or seasons should have been used to obtain parameters in the model. Examples are given below of five such independent tests.

In Fig. 9.3 predicted and measured photosynthetically active radiation (PAR) in a corn canopy are compared. The inputs were direct and diffuse PAR above the canopy, soil PAR reflectance, leaf reflectance and transmittance in the PAR, and LAI.

Fig. 9.4 shows a comparison of the bidirectional reflectance distribution function (BRDF) for soybean, both in the visible (0.4–0.7 μm) and near-infrared (0.75–0.85 μm), between the predictions from Cupid (Norman, Welles & Walter, 1985) and the measurements from Ranson, Biehl & Daughtry (1984). The BRDF, which is used extensively in remote sensing of vegetation, represents how bright the canopy would appear if viewed from various directions by a sensor with a narrow field-of-view (e.g. 5 °). Since the canopy brightness or radiance varies with the view direction, the canopy cannot be represented by an ideal Lambertian scatterer. The inputs to Cupid for this test were soil and leaf spectral properties, LAI, leaf angle distribution and sun and sky irradiance.

In a comparison of predicted water use from Cupid over eight days with measurements, the measured water use was 35 mm and the predicted 29 mm (Norman & Campbell, 1983); a difference well within the errors of measurement. Furthermore,

Fig. 9.3. Comparison of predicted and measured transmitted PAR at two heights (0.19 m and 1.14 m corresponding to *LAI* = 2.5 and *LAI* = 1.4 respectively) in a corn canopy of 0.76 m row spacing at Guelph, Ontario, Canada in 1969. Two days of data are included: 2 Aug. (*, ×) and 3 Aug. (□, ✭). Measurements are averaged over 7 m transects oriented at right-angles to rows. A clumping factor of 0.85 was used in the model to predict the results below (see Nilson, 1971).

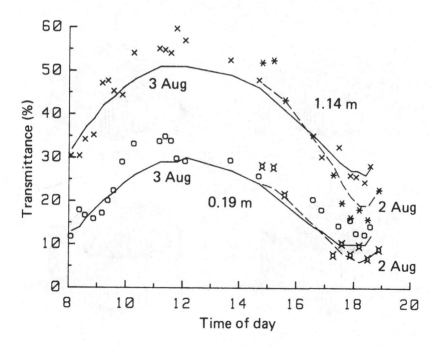

the measured interception of irrigation water during one water application was 3.6 mm ± 2.4 mm and the predicted, 2.8 mm.

Leaf-wetness-duration predictions from Cupid for Wisconsin were compared with measurements of hours of leaf wetness in a crop of Snap Beans over 10 days in July, 1984. Predictions and measurements agreed within one hour with typical durations of 6–12 h. During a second 15-d period in August, nighttime wind speeds were lower and agreement between predictions and measurement were poorer. This emphasises an area of major weakness in our plant–environment knowledge, namely transfer of heat, momentum and mass when wind speeds are low during daytime or nighttime.

The growth of Spider mite populations in corn canopies was compared with predictions from Cupid (Fig. 9.5). In this example Cupid was combined with a dynamic population growth submodel for spider mites (Toole *et al.*, 1984). The mite

Fig. 9.4. Comparison of measured BRDF for soybeans from Ranson *et al.* (1984) with predictions from Cupid (Norman *et al.*, 1985) in the visible and near-infrared parts of the solar spectrum on a clear day. The *LAI* = 3.9, solar zenith = 38 °, solar azimuth = 111 °. The vertical axis is reflectance factor, the centre of the horizontal plane represents the nadir view and radius from the centre is view zenith angle with the outer circle being a view zenith of 60 °. View azimuth is referenced to south being 180 °. The maximum reflectance factor occurs when the sun is behind the observer.

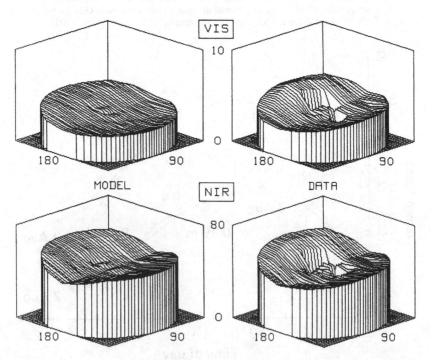

response functions were measured in the laboratory and environmental predictions from Cupid were used as inputs to the mite submodel.

All of the above are independent tests of various aspects of Cupid in that the predicted result was never used in the formulation of the model. These examples demonstrate that reasonable predictions can be obtained over a wide breadth of applications with a detailed plant–environment model. Of course poor agreement between model and measurement should lead to further investigations of the problem.

This is an important part of the discovery process if the rules set forth above are used. However, because you stop when reasonable agreement between model and

Fig. 9.5. Comparison of predicted spider mite population growth with measured growth in corn in central Nebraska as a function of irrigation treatment. The ordinate is the number of banks grass mite female adults per plant. The abscissa is day of the year with 1 Jan. as day 1.

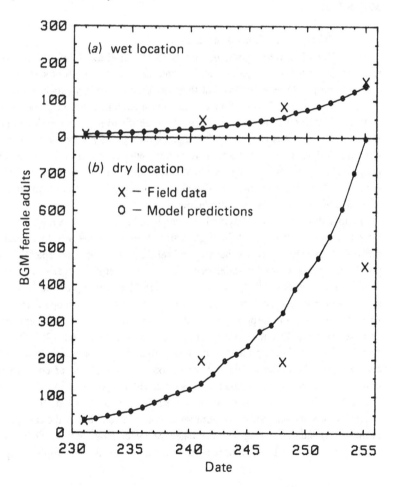

measurement is achieved, there is no guarantee that the explanations are complete or correct, only a reasonable assurance.

Detailed error analysis or sensitivity testing of complex models is extremely difficult. The sensitivity of a given output to errors in some input condition or parameter depends on all the other inputs and parameters. Some reasonable methodologies do exist for attempting senstivity testing of complex models, and the effort is tremendous but probably worthwhile (Huson, 1984). Environmental physiologists are faced with this predicament in measurement programmes as well. The effect of unknown measurement errors, which often are associated with inadequate spatial or temporal sampling, or the effect on interpretation of some quantity that was not measured, contributes uncertainty to experimental conclusions. There is no contest between those who like measurement and those who like modelling; both endeavours help us to tease useful consistencies out of a reluctant Mother Nature.

Practical usefulness as a research tool

Complex, mechanistic models such as Cupid can be extremely useful in research. As our knowledge and understanding of plant–environment interactions expands, no person is so gifted that they can discern the key processes in a given situation. A model such as Cupid can bring a vast amount of accumulated knowledge to bear on a given problem and aid in the identification of the major unknowns. Furthermore, rigorous tests of models against measurements, when that is possible, allows intelligent inference about situations where measurements are not now possible. For example, evaporation or transpiration during rain or irrigation. Mechanistic models certainly can also aid in the design of good experiments. A few examples of such insights are given below.

The irrigation efficiency of agricultural crops in the semi-arid Great Plains of central USA is important because of the high cost of pumping. However, this efficiency is difficult to measure directly because of rain gauge errors and spatial sampling problems. Using a droplet trajectory and droplet energy budget subroutine with Cupid, for a clear day (27 °C air temperature, 18 kPa vapour pressure, 4 m s^{-1} wind speed) over mature corn, more water is predicted to reach the canopy than leaves the sprinkler nozzle. This surprising result occurs because of the low temperature of the irrigation water and the thermal inertia of larger drops (Thompson, 1986). The droplet size distribution of the sprinkler is critical in this analysis, and the farmer has some choice here. The direct cooling of the canopy by application of cold water is as important as the evaporative cooling associated with evaporation from the wet canopy and soil. With this model, one can define precisely what the term efficiency means (inclusion of soil evaporation, evaporation of water intercepted by the canopy, droplet evaporation) and obtain long-term estimates for a wide range of typical environmental conditions. Another interesting insight from this study is that droplet evaporation estimates in the literature are better measures of evaporation from rain gauges than of irrigation water loss.

The duration of leaf wetness is important in the study of disease development. Although condensation has been estimated by weight gains of lysimeters at night when no rain occurs, there can be serious errors. If the atmosphere is not saturated with water and the soil surface is wet, a lysimeter may indicate evaporation (thus implying no leaf wetness) over nighttime hours when perhaps half of that soil water loss is being deposited on the canopy as condensation (Norman & Campbell, 1983).

The temperature differences between air above an irrigated corn canopy and the leaves within that canopy can cause significant errors in the prediction of spider mite development (Toole *et al.*, 1984). Systematic temperature differences of several degrees can persist for weeks. Since the development rates of spider mites may be an exponential function of temperature at temperatures below 35 °C, sizeable errors can occur in mite population predictions if systematic temperature errors persist. Further, model predictions indicate that 1 mm of irrigation every 3 h, something that can be accomplished with centre pivot equipment available today, can retard a mite infestation by a week or more with sizeable economic benefit (Barfield & Norman, 1983).

A model such as Cupid can be useful in combining the engineering-based area of remote sensing with the natural science-based area of plant canopy processes. For example, the ratio of visible-to-near-infrared radiance for a canopy, which can be sensed from satellites (Justice *et al.*, 1985), can be related to intercepted PAR canopy radiative transfer submodels such as those in Cupid (Norman, Welles & Walter, 1985). Moreover, with appropriate leaf photosynthesis submodels, the effect of various factors such as leaf properties, canopy structure, soil properties, sun angle or air temperature on the relationship between satellite observation and canopy photosynthesis can be predicted.

Comprehensive plant–environment models have many other potential applications to research in the natural sciences.

Practical usefulness as a canopy management tool

Comprehensive plant–environment models such as Cupid can indirectly aid management of canopies in several ways. Complex models should aid in the development and testing of simpler models that can be used directly in management decision-making. In addition, the results from studies with detailed models may change management strategies as the knowledge base is expanded. Comprehensive models are not likely to be used directly in management decision-making, at least in being available on-line for the manager to interact with, for some time because of the large program size, long execution times and code maintenance that currently is characteristic of these models. If implementation at the managerial level is ever to occur, and many would question whether this ever should occur, then much more systematic effort will have to be devoted to complex models than is presently the case.

Sustained effort is required to make a sufficiently robust program with inputs and outputs in a suitable form for the non-scientist to use. Computer power is not a limitation, rather scientific knowledge and community motivation are the major obstacles.

References

Barfield, B.J. & Norman, J.M. (1983). Potential for plant-environment modification. In *Plant Production and Management Under Drought Conditions*, eds. J.F. Stone & W.O. Willis. New York: Elsevier Science Publications.

Farquhar, G.D. & von Cammerer, S. (1982). Modelling of photosynthetic response to environmental conditions. in *Encyclopedia of Plant Physiology*, New Series, Vol. 12B. Eds. O.L. Lange, P.S. Nobel, C.B. Osmond & H. Zeigler, pp. 549–88. Berlin: Springer–Verlag.

Farquhar, G.D. & Wong, S.C. (1987). An empirical model of stomatal conductance. *Australian Journal of Plant Physiology*, 11, 191–210.

Goudriaan, J. (1977). *Crop Micrometeorology: A Simulation Study*. Simulation Monographs. Wageningen: Pudoc.

Grace, J., Ford, E.D. & Jarvis, P.G. eds. (1981). *Plants and Their Atmospheric Environment*. London: Blackwell.

Huson, L.W. (1984). Definition and properties of a coefficient of sensitivity for mathematical models. *Ecological Modelling*, 21, 149–59.

Justice, C.O., Townsend, J.R.G., Holben, B.N. & Tucker, C.J. (1985). Analysis of the phenology of global vegetation using meteorological satellite data. *International Journal of Remote Sensing*, 6, 1271–1318.

Nilson, T. (1971). A theoretical analysis of the frequency of gaps in plant stands. *Agricultural Meteorology*, 8, 25–38.

Norman, J.M. (1979). Modelling the complete crop canopy. In *Modification of the Aerial Environment of Crops*, eds. B.J. Barfield & J. Gerber, pp. 249–77. St. Joseph, Michigan: American Society of Agricultural Engineering.

Norman, J.M. (1982). Simulation of microclimates. In *Biometeorology and Integrated Pest Management*, eds. J.L. Hatfield & I.J. Thomason, pp. 65–99. New York: Academic Press.

Norman, J.M. (1986). Instrumentation use in a comprehensive description of plant–environment interactions. In *Advanced Agricultural Instrumentation*, ed. W. Gensler.

Norman, J.M. & Campbell, G.S. (1983). Application of a plant-environment model to problems in irrigation. In *Advances in Irrigation*, ed. D. Hillel, pp. 158–88. New York: Academic Press.

Norman, J.M. & Campbell, G.S. (1988). Canopy structure, In *Physiological Plant Ecology: Field methods and Instrumentation*. eds. J. Ehleringer, H.A. Mooney, R.W. Pearcy & P. Rundel. London: Chapman & Hall.

Norman, J.M., Welles, J.M. & Walter, E.A. (1985). Contrasts among bidirectional reflectance of leaves, canopies and soils. *IEEE Transactions of Geoscience and Remote Sensing*, GE–23, 695–704.

Ranson, K.J., Biehl, L.L. & Daughtry, C.S.T. (1984). *Soybean Canopy Reflectance Modeling Data Sets*. LARS Technical Report 071584. West Lafayette Indiana: Laboratory of Applied Remote Sensing, Purdue University.

Thompson, Allen (1986) Modeling Sprinkler Droplet Evaporation Above a Crop Canopy. PhD Thesis, Department of Agricultural Engineering, University of Nebraska, Lincoln, NE.

Toole, J.L. Norman, J.M., Holtzer, T. & Perring, T. (1984). Simulating banks grass mite population dynamics as a subsystem of a crop canopy-microenvironment model. *Environmental Entomology*, **13**, 329–37.
deWit, C.T. (1965). *Photosynthesis of Leaf Canopies*. Agricultural Research Report 663. Wageningen: Pudoc, The Netherlands..

INDEX

accumulated temperature, *synonym of* thermal
   time q.v.
adaptation, 139–40, 150–1
age structure, 111–13, 116–17, 120–2
air pollution, 36
albedo, *see* reflection coefficient
attenuation coefficient, 24–5
   see also extinction coefficient

biomass, 23, 31, 91
boundary layer; *see also* resistance
   convective, 63–9
   leaf, 165
   planetary, 42
branching, 107, 112, 147–51
   see also tillering

canopy
   closure, 3
   energy balance, *see* energy balance
   influences: nutrition, 83–99, 113–17, 119,
     155; light, 111–13, 129–40; temperature,
     35, 84; water, 35, 88–9, 134–7
   processes, 161–74; *see also* leaf demography,
     evaporation, fluxes, photosynthesis,
     radiation absorption, respiration
   structure, 143–57; description, *see* methods;
     measurement, *see* methods
   temperature, 84
carbon assimilation, 77, 85–7
   see also photosynthesis
conductance, *inverse of* resistance q.v.
convective boundary layer, *see* boundary layer
conversion ratio
   assimilate to dry matter, 29, 88–9
   CO_2 to dry matter, 27
   quanta to CO_2, 27
   see also dry matter: radiation quotient

defoliation, 31, 126
demography, *see* leaf
development, *see* phenology
diaheliotropic movements, *see* solar tracking
dry matter: radiation quotient, 22, 27–31, 35

eddy diffusivity, 45, 52, 57
energy balance
   of canopies, 63, 84
   of leaves, 133–6, 164
entrainment, 66, 71–6
evaporation, 67–9, 71–5, 77–9
   see also transpiration, fluxes
evolution, 149
extinction coefficient of light, 4, 8–9, 137
   see also attenuation coefficient

floral initiation, 107
flowering, *see* reproductive growth
fluxes
   CO_2, 47, 54, 76–7
   heat, 47, 54
   momentum, 43, 47, 54
   water, 43, 47, 54, 63, 67–79

green area index, *see* leaf area index
growth analysis, 21–2
growth rate, *see* productivity

heliotropic movements, *see* solar tracking
herbivory, *see* defoliation

leaf
   angle distribution, 4–7, 34–5, 130, 134,
     137–9
   area density, 3, 5–6, 32–4
   area formation, 36; effect of nitrogen, 93–4
   area index, 4, 25, 32, 135–9, 155–7, 167

Printed in the United States
By Bookmasters